Beiträge zur Kenntnis südasiatischer Sprachen und Literaturen

21

Herausgegeben von Dieter B. Kapp

2010

Harrassowitz Verlag · Wiesbaden

Kāśyapīyakṛṣisūkti

A Sanskrit Work on Agriculture

Edited with an introductory study
by Gyula Wojtilla

2010
Harrassowitz Verlag · Wiesbaden

Bibliografische Information der Deutschen Nationalbibliothek
Die Deutsche Nationalbibliothek verzeichnet diese Publikation in der Deutschen Nationalbibliografie; detaillierte bibliografische Daten sind im Internet über http://dnb.d-nb.de abrufbar.

Bibliographic information published by the Deutsche Nationalbibliothek
The Deutsche Nationalbibliothek lists this publication in the Deutsche Nationalbibliografie; detailed bibliographic data are available in the internet at http://dnb.d-nb.de.

For further information about our publishing program consult our website http://www.harrassowitz-verlag.de

Printed on permanent/durable paper.
Printing and binding: print-z.de, Speyer
Printed in Germany

ISSN 0948-2806
ISBN 978-3-447-06265-7

Contents

Preface

My first encounter with the Kāśyapīyakṛṣisūkti can be dated from the autumn of 1973. At that time I was a postdoctoral research fellow in the Department of Ancient Indian History, Culture and Archaeology at the Benares Hindu University under the supervision of the late Professor Lallanji Gopal the foremost expert in the history of agriculture in early India. Greatly inspired by him and my *bráhmaṇa* friends such as Dr Rajbali Pandey and Dr Ram Adhar Pathak I launched out into an intensive research into various texts concerning ancient and medieval Indian agriculture. Being a bookworm, I enthusiastically visited the numerous bookshops in Benares and one day I happened to come across a book with a very promising title. It was the *Agriculture in ancient India* edited by D. Raghavan and published by the *Indian Council of Agricultural Research*, New Delhi in 1964. This book is a rich mine of information on various aspects of agricultural knowledge. It contains copious excerpts from the Kāśyapīyakṛṣisūkti and among other things a list of manuscripts on *kṛṣiśāstra* (agriculture) and *vṛkṣāyurveda* (gardening) in Indian languages. I had already taken up the study of the running English translation of the text during my stay in Benares in the same year. Simultaneously, I made a firm resolution to study it in Sanskrit original.

It took four years before an opportunity arose to get access to a microfilm copy of the Sanskrit ms. kept at the Adyar Library, Madras (Chennai). It was an immediate result of my visit to the unforgettable Professor V. Raghavan in Madras in November 1977. He was who generously helped me to procure the wanted microfilm copy. It was Mrs. Seetha Neelakanthan, librarian of the Adyar Library who granted me permission to publish the manuscript. The photocopy reached me through official channels of the Government of India in 1978. My first idea was to publish the text in Roman transliteration. I must confess that I was in two minds about whether the edition based on only a single ms. was justifiable. However, the importance of the text for research drove me to put the text before the scholarly world. This text edition was followed by my English translation of the work in 1985.

Since 1977 I have looked in vain for another ms. of the text in vain. I simply failed to trace the original in Tirupati from which the extant transcript had been made. It was a turning point in my work when an edition cum English translation was brought out by the Asian Agri-History Foundation, Secunderabad, Andhra Pradesh, India in 2002 and due to the kindness of Prof. Y.L. Nene, chairman of the Foundation I was sent a copy of this valuable publication within a short period after that. From this book it became me quite clear that the strenuous efforts of the Indian colleagues to obtain another mss. of the Kāśyapīyakṛṣisūkti had also proved to be unsuccessful. At the same time I recognized that, in spite of good solutions of various problems rendered by the Indian colleagues, there still remained a large scope of research into the textual history not to speak of the many philological, semantical, cultural and historical issues.

The next step was to make up my mind to prepare a new edition on the ground of the single extant ms. together with an exhaustive study of the author, date and provenance of the text together with a glossary of special terms exclusively used in the text. Doing so, I could not avoid the problem of the form of the planned edition. At first I thought of a critical edition, but I gave up this idea very soon. The fact that only one ms. existed kept warning me not to opt for this form of edition. On the other hand, it occurred to me that a so-called diplomatic edition could not only be a faithful reproduction of the text but it might serve another purposes, e.g. it might allow us to look into a particular facet of scribal tradition in the Tirupati area where the extant transcript had been made. Finally I decided to publish the text in this form.

During the preparation of this volume I exploited the opinion of a large number of colleagues.

Prof. Oskar von Hinüber (Freiburg, Germany), Prof. Klaus Mylius (Frankfurt am Main), Prof. Thomas Oberlies (Göttingen) and Prof. Chlodwig H. Werba (Wien) all made many valuable suggestions at various points and patiently answered my repeated questions concerning theoretical or technical issues.

In tackling special issues I have greatly benefited from the advices and communications of Dr Gerard Colas (Paris), Prof. Rahul Peter Das (Halle), Dr G. Jan Meulenbeld (Utrecht) and Prof. Jaroslav Vacek (Prague). I wish to express best appreciation and thanks to them.

Last but not least, my thanks are due to Prof. Dieter P. Kapp (Köln) for taking up my work in the series *Beiträge zur Kenntniss südasiatischer Sprachen und Literaturen.* My sincere thanks are also due to *Harrassowitz Verlag* for undertaking the publication of the work.

Budapest 2009 Gy. Wojtilla

Introduction

Author, age, provenance

Author

The treatise has been ascribed to Kāśyapamuni, the sage Kāśyapa, both in the introductory sentence and the colophon. As for the adjectival derivative *Kāśyapīya* that stands in every colophon of the four sections of the text it can be assumed that either Kāśyapa is its traditional author, or at least that this is a work as taught by Kāśyapa.

Arguments can be advanced both for accepting or rejecting his authorship and even the very existence of such a person as Kāśyapa can be questioned.

Kāśyapa turns up as a cultural hero and a promoter of agriculture first in the KA II, 24, 27. J. J. Meyer, who takes here Devala and Kāśyapa as one and the same, thinks that he can be identical with Udalākāśyapa -an otherwise unknown god of agriculture- mentioned in the PāGS II, 13. However, as Meyer himself admits as well that to take the form Udalā as a corruption of Devala meets with some difficulties (Meyer, 1937, 157, n. 1.). Moreover, the reading Devalāya Kāśyapāya stands only in the twelfth century Bhāṣāvyākhyāna commentary on the KA, while the Pratipadapañcikā and Śrīmūla commentaries as well as the most important MSS. of the main text read Devāya (BhāVyā on KA II, 24, 27). On the other hand, the author of the Bhāṣāvyākhyāna might have been thinking of *jyotiṣa* experts such as Kāśyapa and Devala. This Devala can be dated from a period prior to the ninth century (Wojtilla 2006, 67). Consequently he could have not been meant by Kauṭilya or even the later hands working on the extant text of the KA. In this manner together with Kangle I take here Deva and Kāśyapa and Prajāpati as three divine patrons of agriculture (Wojtilla 2005a, 423). However, unlike Kangle (Kangle 1972, 152) I do not associate Deva with rain. I must acknowledge that theoretically Deva or probably *deva* can be an apposition to any of the two names as Meyer assumed (Meyer 1926, 183, n. 2.) or recently Unni (Unni 2006, I, 318) suggests. Whatever the facts may be, these early attestations univocally refer to a mythical person, whose name can be borrowed by anybody to give a great authority to a text concerning agriculture.

Kāśyapa is allocated a similar role in medieval Kashmirian sources such as in the Rt. V, 115) and the NīP 300. Bhaṭṭotpala in his commentary on the Bs LIV, 7 calls Kāśyapa an expert in agricultural meteorology. This later seems to have nothing to do with the author of our text because he does not figure there as an astronomer or climatologist. The existence of a religious text called

Kāśyapajñānakāṇḍa, 'The wisdom-book of Kāśyapa' from the Tirupati area tentatively dated from the second half of the first millennium A.D. (Goudriaan 1965, 10) is also of no use because there is no evidence of a relation between this work and our text (Goudriaan and Colas). Nalini Sadhale ascribes KāKS and Kāśyapadharmasūtra to one and same person (SE, 124). Her statement is not entirely probative because it is founded on incomplete knowledge of the issue. The later text as whole is surely a modern work (Winternitz 1967, 543). There is no evidence of any correlation between these texts and their authors are merely namesakes. Kāśyapa otherwise figures in history as an authority on grammar, medicine, music, poetics, classical prosody and *bhakti* (NCC IV, 144–145).

It is also noteworthy that Kāśyapa calls himself a *muni* and so lists himself among such authorities as Vyāsa the traditional author of MBh, Bharata the author of Nātyaśāstra or the famous triad of grammarians víz., Pāṇini, Kātyāyana and Patañjali (Apte 1957,1278). On the other hand the name Kāśyapa signifies a man belonging to the *Kāśyapa- gotra*, a clan of Brahmins which 'is notorious for being able from early days to absorb (as the name Mātaṅga Kassapa shows) aborigines who wanted to become brahmins'(Kosambi 1975, 118). The distribution of the *Kāśyapa-gotra* between 475–1030 known from epigraphy is instructive: Orissa 6, Gujarat 1, Madhya Pradesh 2, Bengal 2, Uttar Pradesh 1 and South India 1 (S. Datta 1989, 288). The small number in the South is striking.

All in all, we must regard Kāśyapa as a nominal author. He can be neither the author or more precisely the creator nor the compiler of the text. Because the text is regarded as a didactic work in order to give it more prestige is attributed to this semi-god and mythical sage.

It is a puzzling question how Kāśyapa, whose name I use as a shorthand term for the author, exactly worked. He refers to himself as *iti ca proktaṁ kāśyapena maharṣiṇā* 'so said the great saint|||sanctified sage Kāśyapa' (179), or *śāstre 'sminn api kāśyape* 'also in this textbook of Kāśyapa'(verse 536) ,or he uses the phrase *mataṁ mama* 'my opinion'(verse 6), or he uses the expression *iti matir mama* 'this is my opinion' (verse 607) because such devices are employed in *śāstric* literature without a compulsory, real author. All this may simply indicate that he had some predecessors and he does not agree on their statements. Unfortunately, he does not give the names of the authors of these statements. Unlike to Surapāla or Parāśara, the authors of the Vā. and the KṛP respectively, he does not take care of any concord of contents and form. He did not employ special metres if they could better fit the contents. Composing sentences he frequently transgresses the verse limits.

Had he sanskritized vernacular sayings or verses he would have kept sentence limits or verse limits.Verses 681–682 lend some support to the Sanskritization hypothesis. 'Then the *kṛṣiśāstra* related by the [Goddess] Earth (Bhūmi= Bhūmidevī) was brought to Earth by the sages. It was preserved by the kings who bear the burden of the protection of their subjects and then in the course of time it was particularly studied by the *śūdras*. Then the *kṛṣiśāstra* was greatly extended by the inhabitants of the Earth according to the time of rainfall.'

Or, does the extant text bear the touch of the editorial work of more hands i.e. less educated editors copyyists? One can easily single out instances of symptoms of the loose edition of the text. The fact that Kāśyapa the agriculturist has not been quoted in any Sanskrit text known to me seems to advocate the later hypothesis.

In the light of the tradition he is a nominal author in much the same way as Kauṭilya of the KA or Manu of the MāS.

Age or ages

Sanskrit texts are notoriously difficult to date and our treatise exemplifies this vexed problem too. Randhawa proposes a date from the 5th and 10th centuries A.D. (Randhawa 1980, 484). Nene thinks of a date between c. 700–800 A.D. (SE, 132). The above discussion on Kāśyapa's person might allow us to think tentatively of the early medieval period. However, this is only one aspect of the problem. Therefore we have to look for further evidence.

Total lack of testimonia

The KāKS has not been cited in any known Sanskrit text. The twelfth century Pratipadapañcikā of Bhaṭṭasvāmin that comments upon numerous passages of the KA concerning agricultural production mentions only Parāśara as an authority on agricultural science, but keeps silent about Kāśyapa. More disturbing is the fact that the modern commentary on the same work called Śrīmūla written by T. Ganapati Sastri, a southerner, does not know Kāśyapa either.

Direct references to possible sources i.e. former authorities

On the other hand there are few references to former authorities on various subjects dealt with in the text. Verses 523–524 say that ' In this world various foods are made from rice which comes from the paddy for the sake of protecting life. The *pākaśāstraka* gives instruction concerning this matter. King Vīrasena composed a big *pākaśāstra* and his son Nala and the great sage Garga did likewise.' Vīrasena looks like a misnomer for Bhīmasena i.e. Bhīma who disguised himself as a cook of soups sauces (*sūpakartṛ*: MBh IV, 2, 7) and who, according to the ninth century Bṛhatkathāślokasaṁgraha of Budhasvāmin (BKS XVI, 61 and XVIII, 20), was an expert in *sūdaśāstra* 'the science of cooking'. Nala, alias Bāhuka, served Ṛtuparṇa, the king of Ayodhyā, as his chief charioteer and also as his cook. We learn from the Nala story (MBh III, 73, 10–12) how once he prepared a meat dish for Ṛtuparṇa. Somadeva shows him as a miraculous cook (KSS IX, 6, 395–96). Nevertheless, as Basham justly put it, 'no ancient textbook on cookery has survived' (Basham 1981, 215–216).

At the same time, the extant work under Nala's name, called Pākadarpaṇa, 'The Mirror of Cooking' is a relatively late text on Indian dietetics. Since it does not mention chili (red pepper) and other vegetable stuff introduced by the Portugese, it may have been composed before the sixteenth century, but this inference is not certain (Das 2001, 121–122). As I see it, the book is clearly a text that belongs to that branch of *āyurveda* which became popular after the sixteenth century (Meulenbeld 1984, 66 n. 70 and Wojtilla 2004, 339). This kind of cooking was no longer practised by professionals other than kings. The wording of the Pd. concides with that of the MBh as far as its anonymous author speaks of the preparation of *sūpa* from pulse and vegetables (Pd. pp. 11 - 12). However, this account significantly differs from that of the MBh. This texts prescribes a healthy soup while the latter refers to a delicious one. At the same time it is noteworthy that at the commence of the Pd. beside the here usual god Gaṇeśa Veṅkateśa and Dhanvantari are invoked. The first one is known as the physician of gods in Hindu mythology and the traditional author of a medico-botanical glossary called Dhanvantarinighaṇṭu. Veṅkateśa is a form of Viṣṇu worshipped in Tirumalai, near Tirupati. It can be conjectured that at a certain time both the KāKS and the Pd. had simultaneously gained currency in the Tirupati area.

The thirteenth century commentary called Jayamaṅgalā on the Kāmasūtra (Jm. I, 3, 15) says that the sauce *(vyañjana)* made of vegetables makes food especially tasty. Kāśyapa uses also the term *vyañjana* (verses 522 and 525) in the same context but not *sūpa.*

It is not without interest that there is an unpublished text of the genre of the Pd. and it is ascribed to Bhīmasena (Wojtilla 2004, 344).

These data allow us to make some tentative assertions. First of all, it is clear that here we have to do with a strong tradition of attributing books on traditional sciences to gods, demigods or mythical persons. This practice works in the case of the above cookery books. It stands to reason to think that in verses 523–524 these books are meant. Being so, the dating of the Pākadarpaṇa of Nala may be helpful for the dating of our treatise to a period around the sixteenth century. Further research into the textual history of the Pākadarpaṇa and Bhīmasena's book may refine the dating. Of course, it calls for further explanation of why in the above verses stand *Pākaśāstraka* instead of Pākadarpaṇa and Vīrasena for Bhīmasena. It can simply be due to the contamination of names of persons and books ascribed to them. All in all, I find the coincidence of wording in our treatise and the Jm. a strong argument for the thirteenth century as *post quem* in dating.

The name of sage Bhārgava occurs in the text in three times.In the context of fruitful cooperation in agricultural activity verse 218 says 'Simultaneous uttering and unanimity is seen everywhere with the bees – so declared in this world Bhārgava the sage.' Verse 311 reads 'From the protection of cows the highest pleasure is born for gods and therefore welfare for the subjects – so said Bhārgava.' While Bhārgava recommends table-land for growing *āḍhaka* (pigeon pea, *Cajanus indicus* Spreng.) and *māṣa* (bean marked with black and grey spots,

Phaseolus mungo Linn.) (verse 535), Kāśyapa says that *māṣa* yields also fruit in low places in the rice-field. This Bhārgava seems to be the person who is said to be the expounder of Nītiśāstra (MahāBhā XII, 210, 20) and certain verses are even ascribed to him (MahāBhā XII, 57, 40–41 and 139, 70–71 cf. Kane 1968, I, 265). It is an intricate question as to whether Bhārgava, the authority cited three times, has anything to do with Bhārgava Rāma alias Paraśurāma who instructed Kerala *brāhmaṇas* on agriculture in the Malayalam Kṛṣigītā (KṛṣiGī Chapter One).

Verse 553 says '[Agricultural activity] is described as life-giving to cows, bulls, camels, horses, elephants and others by the best of *yogis* beginning with Nārada.' This appears in a passage quoted from an unknown source. At least I have failed to trace it. To call Nārada a *yogi* is uncommon.The Mānavadharmaśāstra (ManuSm I, 34–35) lists him among the ten great sages (*maharṣi*). In the Nāradasmṛti proper there is only one passage (Nārasm XI, 19) which is concerned with grain production. It says that grain cannot be produced without water, but too much water spoils it. Both too much water and insufficient water are injurious to grains.

Verse 64 reads thus. ' The fertile land of villages, towns and cities is said to be fit for grain, which is the breadth of life, the life, the best thing, the artery of life – so said Gārgya the foremost among the authorities.' Gārgya, also known asTrijaṭa in the Rāmāyaṇa (fourth century B.C- fourth century A.D.), was a sage living by farming. Being a poor man with many children, he asked Rāma for charity. Thereupon he was miraculously donated one hundred thousand cattles (Rāmā II, 32, 29–39). Gārgya is also quoted in the Kṛṣiparāśara where two verses (KṛṣPa 164–167) are attributed to him. The relevant verses concern with the careful preservation of seeds before sowing. The upper time limit of the compilation of the later is the eleventh century A.D.

Indirect references to datable texts

The list of foods or food ingredients the *brāhmaṇas* are prohibited from is the subject matter of chapter three. A close examination of its contents shows that Kāśyapa heavily draws on the MaS and the VaiSmS. The first one can be dated from the period between the 2nd century BC and AD 2nd century (Olivelle 2004, XIII), the second one from the AD 3rd-4th century (Caland 1929, XVI and Gonda 1977, 143).

Accidental indications of various ages

According to Nene the examination of the soil before digging wells (verses 48–53) is a procedure 'very similar to what is done today' (SE, 134). Certain elements of the examination based on smell, taste and colour can be attested in other sources: the twelfth century Vṛkṣāyurveda of Sūrapāla (Vā. 47–53) refers to the tests of colour and taste; the fourteenth century Upavanavinoda (Uv. X) mentions the examination of the taste of the soil; the test in the Viśvavallabha, a text written about 1577, relies on colour and taste (Vv. III, 2–3).

Seeds were stored among others in vessels made of strong glass (*dr̥dhakāca* in the verse 494). The beginnings of glass industry may go back to the first half of the first millennium BC (Rau 1983, 10). However, its production was restricted to manufacturing tit-bits like beads and bangles (Sangwan 1990, 76.) Prakrit *kayavatra* can tentatively be derived from Sanskrit *kācapātra* glass vessel in a *kharoṣṭhī* document from Nīya (No. 534) dated from the 3rd century A.D. (Burrow 1937, 81) where it stands on a list of deposited goods like cloth, bow and arrows, a *pothi,* one rug, one carpet, one felt-garment, ropes and some unidentified things may indicate that it was a household utensil in that area. Due to the close commercial connections with India and the possible derivation of the Prakrit name it can be an indirect proof of the existence of the same type of vessel in India too. In want of contemporary linguistic data and archaeological evidence from India, however, it remains only a hypothesis. According to certain Sanskrit sources glass utensils were used in pharmacy as early as the 13th century A.D. (Gode 1969, 93), but the archaeological evidence is lacking. It looks like that 'vessels of glass began to be manufactured during the Muslim rule' and then a great development took place in the 17th-18th centuries under European influence (Sangwan 1990, 76).

The *śaṅkara*-type well is referred to in verse 428. It can also be attested in the twelfth-thirteenth century Aparājitapr̥cchā (Ap. p.185) and the Vv. II, 14.

The distinguished position of the *śūdras* in the rural society depicted by the text deserves attention. It is stated that *kr̥ṣiśāstra* 'in the course of time was particularly studied by the *śūdras*' (verse 681) and 'The practice of agriculture – a measure for welfare and victory derived from sages everywhere – in some places is confined to the *śūdras* too' (verse 208) and '*śūdras* as hired workers (*bhr̥tya*) are known in the villages' (verse 211) or employed in harvesting grain (verse 481). This situation may fit the transitional period when *śūdras* besides being slaves and hired workers came up as peasants whereas 'it became necessary to call them cultivators' (Sharma 1980, 257).

The occurence of the social term *anujīvika* 'dependent of a tenant cultivator', ' a dependant' in verse 480 compels attention. For in verse 319 *bhr̥tya* and *sevaka* definitely denote 'hired labourer and''servant' the exact social position of them is not quite clear. This form cannot be attested in any Sanskrit text. The closest word to it is *anujīvana* '(royal) servant' which stand in the PaŚR 870, a text datable from the period before AD 1200. The corresponding Prakrit form without the diminutive suffix –ka, namely *anujīvi,* stands in the meaning ' who has to depend (on somebody) for subsistence or support, a servant, an attendand' CCDPL 303). The Pāli *anujīvin* is also 'a dependant' (PTSD 35).

Verse 155 says that the king should employ brave soldiers (*bhaṭa*) in the work of digging wells. *Bhaṭas* were soldiers of the king with the function of maintaining peace in villages (Nyammasch 1984, 99 and Choudhary 1971, 117).

Normally they were not welcome there and it was a usual royal privilage to ensure villages to be free from them (Nyammasch 1984, 102). In short, the above job *bhaṭas* is exceptional and not recorded in any other Sanskrit text. Much less probably, being a Prākrit form of Sanskrit *bhṛtya* it stands here in the sense of hired-labourer.

Some consideration should be given to verses 491–492 which follow the work of harvesting, threshing and winnowing grains. These verses speak thus, 'Having given one part as a gift to the Gods and the kings according to the rule and one part to the *brāhmaṇa* learned in the Vedas according to the rule and one for nourishing the hired labourers, he should store the remaining stock of paddy in his house.' It may be the earliest clear reference to the *jajmānī*-system (Sharma). It is generally held that the more distinct outlines of this system may be detected around the 10th century (Nyammasch 1984, 135). Notwithstanding, traces of its origins may be attested in the great Epics (Kölver 1988, 281).

The strong royal patronage of agriculture is a salient feature of the whole text. Verse 626 explicitly says, 'The *kṣatriya* king, who bears the burden of protecting his subjects, should protect agricultural work which is praised as the superb activity among all occupations.' The verses 766–768 go into the details, ' The king, having known the art of cultivation and the art of sowing seeds according to the season and time and the agricultural work which is dependent on water-reservoirs and rain water, should manage the agricultural work in accordance with the fertility of the soil.' Some of these ideas can be read in the Bārhaspatya Arthaśāstra, datable to the period before 700 AD and the ninth century Mahāpurāṇa (cf. Singh 1968, 68).

Verse 205 specifies people who can be nominated for a membership in the *grāmanyāyasabhā*. This term is unknown in other Sanskrit texts. It is likely a village council controlling the irrigational system, safeguarding people against robbers and maintaining right manners (*nyāya*). The existence of such assemblies is a distinctive feature of South India between the ninth and fourteenth centuries. These local bodies had a primacy in allotting agricultural resources (Stein 1994, 47).

There are instances that make us suspicious of very late interpolations to the MS. The occurrence of certain sweets and particularly of *jalebi* in section four may indicate an age extending to 1500 - 1600 (cf. Meulenbeld 1984, 66, n. 70 and Gode 1943, 170 and 173). Mahias opines that all sweets were known before the 10th century A.D. (Mahias 1988, 282). Randhawa' remark 'observations of Kashyapa on the cultivation of rice are remarkable for their accuracy' (Randhawa 1980, 485) holds good as far it goes. However, the prolonged research may give us a hint at modernity. The great number of species of rice (verses 362–368) is unparalleled in Sanskrit or even Bengali literature and the description of the ripening process of the rice (verses 460–466) looks like the work of a modern observer. Similarly the extraordinary great number of *termini technici* coined by help of the word *kṛṣi* otherwise not to be found in such a quantity or quality in any other extant kṛṣiśāstra may make us suspect modernity. The existence of modern *śāstras* borrowing material of nineteenth century scientific literature such as the Śukranīti or the Vṛkṣāyurveda of Parāśara cannot totally be overlooked (cf. Gopal 1962, 524 and recently Sharma 2003, 17–19). Notwithstanding, I do not regard the presence of quasi-modernism as dominant as those in the Śukranīti (Wojtilla 1993, 274). Verses 681–683 can be understood in a twofold way: agricultural lore reached common cultivators through holy and learned mediators; the divine wisdom of agriculture was tarnslated into vernacular languages. In the field of the climatological science which is intimately connected with agriculture there are instances of rendering pieces of Sanskrit texts by pundits into vernaculars (Tripathi 1971, 6). The appearance of such terms as *maṅgalasūtra* n. 'sacred thread worn by married women on their neck' make us cautious. As R. Pandey puts it the name 'is not mentioned in the Gṛhyasūtras and its origin seems to be more popular than scriptural' (Pandey 1969, 216). The word cannot be attested in any traditional Sanskrit dictionary (*kośa*). It is recorded in Hindi dictionaries where it means ' the thread wrapped round the wright wrist as an emblem of a god's gift"(Bhargava 1973, 836) and also the „thread that connects the hands of the bride and bridegroom at the marriage ceremony" or „thread worn by married women on their neck"(Beskrovniy 1972, II, 318). It appears as *mangalasūtramu* n. „the cord or necklace fastened on the neck of the bride as a token of marriage" (Brown, 1980, 938). It is likely that the Telugu word , which is of Sanskrit origin, inspired the editor or the scribe of the extant text of the KāKS.

Speaking of the Sanskrit works attributed to Kāśyapa the case of the Kāśyapasaṁhitā or Vṛddhajīvakīya Tantra is also edifying. This text which contains a valuable chapter, called Bhojanakalpādhyāya (KāS pp. 168–176), on the preparation of fluid medicaments is an example of such works that are 'recent adaptations of earlier texts, and as a rule they are simply modern works that are decorated with ancient names' (Winternitz, 1967, 627). Some scholars date it from a period before the 7th century AD (EDSHP LXXVIII). R. C. Majumdar's idea is rather perplexing. He thinks that the text is 'of an age not much later than that of the Bhela Saṁhitā' while dates the 'original Bhela Saṁhitā' from the period of the Brāhmaṇas (Majumdar 1971, 227 and 222).

In short, the bulk of the text can be dated from the medieval period, but there are passages from a much later time and even quite modern interpolations. On the basis of the single manuscript these cannot be satisfactorily separated.

Provenance

As to the provenance of the work we have the floor for a great deal of questions. From the various data a complex picture emerges which aptly illustrates the textual history of the treatise. The data which can be involved in search for a plausible provenance broadly fall into three groups: some may point to North India, others to South India and again others to an all-Indian situation.

North India

The comparison of the two lists of places in India where agricultural production is going on is very informative While on the first list stand the names Kāśmīra, Vaṅga, Nepāla, Pañcāla, Kosala, Kuru, Virāṭa, Avanti, Mālava, Śakadeśya, Sindhu-Sauvīra, Śūrasena, Cedi, Koṅkāna and Andhra (verses 321–322) in the second part we meet and another list comprising Gāndhāra, Kunti, Pāñcāla, Kāśmīra, Avanti, Sindhu, Nepāla, Naiṣādha, Kosala, Aṅga, Ghūrjara and Saurāṣṭra (verses 751–752). A cursory glance on them reveals that some names stand on both lists: Avanti, Kāśmīra, Kosala, Nepāla Pañcāla, and Sindhu-. These names refer to North and Central India. Moreover the list one covers also the same area. The list two provides some hints to territories lying slightly southwards from Central India. The real South is wanting.

The above lists of country names delineate a territory of the Indian subcontinent that embraces the whole North and Central India approximately up to the southern end of Andhra Pradesh whereas the real South India is totally missing from it.

The instruction how to keep animals in the hot and the cold season respectively (verse 316) may speak in favour of North India.

The appearance of the camel (*uṣṭra*) in verses 353 and 803 - which is a pack-animal of the arid areas of the subcontinent and mostly bred in present day Rajasthan, Gujarat, Haryana and Ladakh- may advocate North India. This argument may lose from its weight if we take into account the fact that these verses are quotations from certain authorities, the sage Nārada and the VaiSmS IX, 15 and it belongs to the all-Indian information mine of the author or the „editor editors.

Randhawa highlights the importance of irrigation in the system of cultivation propagated by Kāśyapa and takes Kosala víz., part of Uttar Pradesh as the place of provenance (Randhawa 1980, 484). Verse 424 reads thus: ' The sages have determined that rice cultivation is best in countries like Kosala.' It is true, the geographical name Kosala occurs in both above lists of the countries where cultivation of land is going on. Beside Uttar Pradesh, Bihar, Bengal, Orissa and Madhya Pradesh are the major rice producing areas. A further argument for Kosala is that this country is called *adevamātr̥ka* 'not dependent on rain for water i.e. watered by rivers, tanks and canals (Rām. II, 100, 45). Kosala is a joyful, prosperous, great country located on the banks of the river Sarayū (*niviṣṭaḥ sarayūtīre*) and she possesses plenty of wealth and grain (Rām. I, 5, 5). Bengal and Orissa has plenty of rain. These countries do not need artificial irrigation: they are *devamātr̥ka* territories. The famous agricultural treatise the Kr̥P which had gained its recent form in the eleventh century (Gopal 1973, 168) does not mention any kind of artificial irrigation. The major part of the work deals with climatology. Although the text became later popular in other parts of north India its provenience is most likely North Bengal (Gopal 1973, 153–154).

South India

Nene suggests with a considerable argumentation that the main focus of Kāśyapa was the Krishna – Godavari deltas and the adjacent areas: the region has assured rainfall to fill up reservoirs; clayely red to black soil; the basic breed is Ongole breed; the terms śambaka and kalama are still used for rice types in southern India; at least 80 plant species, of the 125, exist today in the state of Andhra Pradesh; the technique of transplanting rice was widely practiced in Krishna – Godavari deltas in 100 A.D (SE 146). As a matter of fact the single extant manuscript, a transcript made in Tirupati, comes from the Tamil speaking area of Andhra Pradesh. The above-mentioned developments in the Pallava reign, the construction of large reservoirs in the seventh-eleventh centuries (Adiceam 1966, 164–166) and afterwards up to ca. 1300 A.D (Myrdal –Runebjer 1988, 19–21) imply a south Indian provenience of the extant work. The employment of elephants in working water-lifting contrivances (verse 167) looks like a south Indian practice. The sixteenth- eighteenth century Sanskrit text called Kr̥ṣisamayanirṇaya shows that the elephants play a very important role in Kerala in everyday life (Wojtilla 2005b, 376).

We are told that Bhūdevī was supported by Varāha, the boar incarnation of Viṣṇu. Although already the Rām. describes the origin of Earth and the uplifting of the Earth by Viṣṇu and then the Viṣṇupurāṇa, Liṅgapurāṇa and the Garuḍapurāṇa mention Brahmā as the up lifter of the Earth from the ocean. Brahmā is everywhere identified with Viṣṇu (Rao 1985, 128–129). The myth appears as a beloved topic in fine arts: the representations in Mahābalipuram and Bādāmi are from the early Chālukya-Pallava period (Rao 1985, I, 1, 137–141 and plates XXXVI and XXXVII). The late Pallava rulers who followed Vaiṣṇava faith gave strong patronage to paddy cultivation and the Tirupati area – the place to where the provenance of the single manuscript of the text can be traced- roughly belongs to this region. This region is the home of the Vaikhānasa Viṣṇu rites in the early medieval period. As Gonda puts it the worship as practised by Vaikhānasas was embedded in agricultural environment and it was intimately connected with the welfare of the community, of the village, the successful growth of the crops and the like (Gonda 1992, 236).There is a Tamil record of 1235 from the shrine Śrī Govindarāja in Tirupati isued by Bhūdevīputra-Citrameḻa for the maintenance of the *dharma* observed by the four *varṇas* (Stein 1994, 222). In an inscription written in Sanskrit and in Tamil originated from Tirukkoyilur, South Arcot (Tamil nadu) and dated approximately from the first quarter of the 13th century Bhūmidevī was the beneficiary of a grant. As the inscription says this order (*śāsana*) is 'for the protection and strengthening of the sons of the land, who are born of the four castes etc.' (Stein 1994, 220). Incidentally our text speaks also of people belonging to the four castes whose livelihood is secured by the Goddess Earth.

All-India

The geographical distribution of plants (SE 151–158) is of little help, it points to the whole of the Indian subcontinent. This circumstance is well in harmony with the all Indian character of the majority of classical Sanskrit literature where the authors, even if their place of birth or activity can roughly, boast of their universal knowledge of India and her tradition. Towards the same direction show some universal religious ideas in the work. The whole text reflects a Vaiṣṇava *Weltanschauung* in which Viṣṇu and his consort stand with uniquely numerous names (Jagannātha, Kṛṣṇa, Paramātmā, Śārṅgin, Śārṅgipara and Bhūdevī, Devī, Dhānyadevī, Gaurī, Lakṣmī,. Nevertheless the verse 247 declares that a wise man, who knows agriculture and keen-witted, before starting with agricultural operations, should first meditate upon the divine Trimūrti. The idea of the divine Trinity appears first in Kālidāsa's Kumārasasambhava (Ks. VII, 44, cf. Glasenapp 1922, 141): 'The one only form was divided three fold. Sometimes Hara or Siva was senior to Vishnu and Hari or Vishnu to him and Brahma to both; and they too Vishnu and Siva were sometimes prior to Brahma'. (H. H. Wilson's translation). The name Trimūrti in Ks. II, 4 seem to refer only to Brahmā. It is agreed that this name as a common term is first mentioned in the thirteenth century Caturvargacintāmaṇi of Hemacandra (cf. MW 460). The divine Trinity is depicted in a nice Mughal miniature from circa 1600 (cf. Randhawa 1982, Fig. 34) where farmers are seen threshing and winnowing grain and above them the persons of the Hindu trinity, Viṣṇu, Brahmā and Śiva seating on clouds.

The term *ṛsyajña* 'sacrifice to the sages, i.e. study of the Veda' deserves also attention because it occurs only in the MaS IV, 21 (cf. MW 227).

These items also bear the testimony of a kind of universality, since the Mānavadharmaśāstra has been the foremost code of law in India.

Quite safely it must be said that the text in its extent form might be the work of a south Indian editor who freely used the rich proverbial literature (*sūkti*) from different parts of India. The contents of verses 682–683 can be conceived as editorial notes: 'the *kṛṣiśāstra* was greatly extended by the inhabitants of the Earth according to the time of rainfall; in this manner people diligently studied their own *kṛṣiśāstra.*' It must be admitted that one of his major sources must have come from the historical Kosala, the present day Uttar Pradesh, and another Tamil Nadu: the repeated stress on the former location may reflect a very old tradition on the origin of rice cultivation in the north; the long passages depicting seemingly southern situation may be due to conditions prevailing in the time of final redaction of the text.

The structure of the work

Our text is a treatise on agriculture and related subjects, on royal duties in the organization of rural society and preparation of ritually pure food to be eaten by the *brāhmaṇas* and used for making offerings to gods.

As the title says the work is a collection of wise sayings (*sūkti*) and as such it somehow

follows a pattern in its structure reflected in *kṛṣiśāstra* texts composed either in Sanskrit or in vernaculars. Although it contains chapters devoted to certain topics there are a great deal of passages which do not fit into the given context or which reappear in a slightly modified form in an other chapter. The traditional history of the beginnings of agriculture on Earth is first encapsulated in the story of Viṣṇu's *Varāha avatāra* (verses 9–23) and there is an another version not mentioning Viṣṇu at all but exclusively connecting it with Brahmā (verses 614–615 and 678–680). A brief *arthaśāstra* is embedded in part two (verses 683–777, cf. Wojtilla 2005c, 343–351) where some main issues lengthy discussed above are revisited. It is to surprise that the operations beginning with harvesting and ending with winnowing is described at great length first in verses 479–490 and then once more in an abridged form and using some particular terms in verses 561–565. For there is only one extant manuscript which is itself a special transcript it would be haphazard to reconstruct an 'original' text with omitting the alleged interpolations or evident repetitions. The loosely edited text, as it stands, contains four sections (*bhāga*).

Section one, called 'The Description of the Methods of Cultivating Grains etc.' (*dhānyākṛṣikramakathanam*), consists of fourteen chapters including an introductory chapter called 'Preliminary Advice on the Subject' (*śāstropadeśakrama*). This introduction highlights the importance of agriculture for maintaining traditional Hindu society and gives a Vaiṣṇava myth on the origin of the grain-producing Earth. The other chapters are as follows.

2. The Description of the Methods of Dividing the Soil According to its Fitness for Producing Grain into the Following Types: Hilly Land, Land Irrigated by Rivers and Streams, Forest-land Pasture-land and the like (*girinadīnadavanasārakṣetrādibhedena sasyārhabhūmivibhajanakramakathanam*)
3. The Description of the Methods of Establishing of Water-reservoirs on the Land (*bhūmyāṁ jalādhārasthāpanakramakathanam*
4. The Description of the Methods [Establishing] Canals Depending on Rivers, Wells and the like (*nadīmātṛkakulyākūpādikramakathanam*)

5. The Description of the Characteristic Marks of Cultivators and the Village Officers and the Eulogy of Agriculture (*kṛṣivalānāṁ grāmakāryakarāṇāṁ ca guṇalakṣaṇakathanam* and *kṛṣipraśaṁsā*)
6. The Description of the Collection of the Requisites for Agriculture (*kṛṣikāryārhadravyasaṁgrahaṇakathanam*)
7. The Description of the Method of Plough Worship (*lāṅgalapūjākramakathanam*)
8. The Description of the Method of Worshipping the Bull-king (*vṛṣabharājapūjākramakathanam*)
9. The Methods of Descripting the Marks of the Cows and Bulls Completely(*saṁgraheṇa govṛṣabhalakṣaṇakathanakramaḥ*)
10. The Method of Investigation of the Proper Season for Grains and the Methods How to Divide the Land (*sasyārhakālaparīkṣākramaḥ kṣetravibhāgakramaśca*)
11. The Description of the Methods How to Collect Various Seeds (*nānāvidhibījasaṁgrahaṇakramakathanam*)
12. The Methods of Ploughing for the Purpose of Sowing (*bījavāpārthaṁ bhūkarṣaṇakramaḥ*)
13. The Description of the Methods of Cultivation Which are Suitable for Obtaining Various Grains (*vividhadhānyaprāptyarthaṁ kṛṣikramādikathanam*)
15. The Description of Agricultural Methods in Order to Obtain Various Substances Beginning with Pulses (*āḍhakādivividhadravyaprāptyarthaṁ kṛṣikramādivarṇanam*).

It must be noted that due to the loose redaction of the extant text there are some peculiarities: chapters 1–9 are included into the *śāstropadeśakramaḥ* and the *dhānyādikṛṣikramakathanam* proper begins with chapter 10; chapter 13 embraces almost all the topics discussed in the other chapters and so it forms a quasi-independent treatise. Moreover this chapter definitely highlights Kosala as the best grain-producing place in India while chapter 10 speaks of same activity on an all- Indian scale. Chapter 14 discusses various food stuffs beginning with pulses with a special emphasis: these products serve as ingredients of different foods and especially of sauces. The way how to of prepare the later can be learnt from the textbooks of cooking (*pākaśāstra*) authored by Vīrasena and Nala.

Section two, called 'The Description of the Methods of Growing Vegetables etc.' (*śākādikṛṣkramakathanam*), is of heterogeneous contents. In accordance with its title it deals with the cultivation of vegetables, fruits and trees but beside that it gives room to a minor *arthaśāstra*, a textbook on economics and politics (683–777 cf. Wojtilla 2005a), which treats all spheres of economic life mainly based on agriculture. Forestation ranks among the main tasks of the kings and accordingly it has been encouraged to be carried out in a series of countries listed here.

Section three called 'The Description of Rules on Edible and Inedible Things' (*bhojyābhojyakramakathanam*) and Section four called 'The Description of the

Order of Making Various Oblations' (*vividhahavyanivedanakramakathanam*) are regarded as appendices by Nalini Sadhale (SE, 129). In the light of the introductory part of chapter 14 and the in social-religious ideas frequently emphasized in this treatise I take section three and four more closely connected with the other parts of the text: the whole work represents an ideal society living on prosperous agriculture where cultivation of plants plays dominant part and where vegetarian dishes are propagated while non-vegetarian ones are only silently permitted without a real discussion of meat production or the like.

Language and style

The language of Kāśyapīyakr̥ṣisūkti is very simple. It resembles the modest sentence forms and style of certain portions of purāṇas and *śāstric* literature. Special verbal forms are absent: the commonest forms are present indicative, present optative and in special constructions passive participles while imperfect or perfect forms are seldom used. In few cases the usage of voice does not follow the rules of classical Sanskrit (obviously for reason of metre stands in 15b *modati* instead of *modate*). There are deviations from concord in number such as plural for singular (*bhūpo vardhayeyur* for *bhūpo vardhayet*); in genre (*śreṇikā dr̥ṣṭaḥ* for *śreṇikā dr̥ṣṭā, tatpūrtiḥ dr̥ṣṭaḥ* for *tatpūrtiḥ dr̥ṣṭā, sasyādivr̥ddhis tatraiva niścitaḥ sukhadāyakaḥ* for *sasyādivr̥dhis tatraiva niścitā sukhadāyakā*). Deviations from *sandhi* are seldom (816b *śarkarādyaiḥ guḍair* for *śarkarādyair guḍair*). For the sake of exactness forms made from nouns with the verb *kr̥-* are not infrequent. For the same purpose determinative compounds (*tatpuruṣa*) were used in great number. Such forms as *śālikṣetr*a, *śālibhū, śālibhūmi* and *śāleyabhūmi* are generally used instead of *śāleya* (P V, 2, 2) or *ājaka* beside *ajāśakr̥t* or *ākrīḍodyāna* instead of simple *ākrīḍa* or *udyāna* primarily for semantical or metrical reasons. A striking feature is the abundance of compounds consisting of more than two members (cf. *janabāhulyanagararājadhānyādivāstuka* 'an overpopulated settlement such as a town, a capital and the like'). This phenomenon could be explained for signs of a relative modern age of the composition of the work. However, such an explanation is not waterproof: there are long compounds in the philosophical texts, or in special scientific texts such as the Vā., not to mention instances even in classical Sanskrit poetry. Fine examples can be quoted from quite different texts: *analpaguḍānvitāmbusikta* 'sprinkled with water mixed with plenty of honey' (Vā. V 57); *syūta-palāśa-patra-puṭikā-pātra* 'a hollow bag made of sewn *palásha* leaves' (Vś. 93, G. Bailey's translation); *akr̥ṣṭoptā taṇḍulaphalaśāli* ' lit. rice growing without ploughing and sowing and producing grain' (Mav. 5310). Compounding is generally a very productive means of coining special terms. A term such as *kr̥ṣikṣetra* 'agricultural land' is quite exact while *kṣetra* is a polysemous word. Such a precise term as *tārabhūkṣetra* 'a field situated on table land' is a credit to the author. Notwithstanding, compounds do not mean automatically exactness. Nalini Sadhale, perhaps somehow rigidly, says that 'conciseness and precision of expression which are merits of a writer on the subject of science are also missing here. Examples of prolixity and verbosity are in plenty' (Se 131). Against this severe judgment I may bring up the special subject-matter and occasionally the lack of poetical genius of our compiler(s) as mitigating circumstances.

The main reason for that may be that unlike the Kr̥P, a genuine collection of sayings (*sūkti*) strictly meant for oral transmission, the KāKS or the Vā. are more theoretical, most likely meant for reading. At any rate, to label the single passages saying and the whole work a collection of sayings is somehow ambiguous because frequent reference to the *śāstric* knowledge in our text seems to contradict to this assumption. With other words our text does not fall in the category of the vernacular sayings on agriculture such as the Ḍāker-bocan, Ghāgh aur Bhaḍḍarī kī Kahāvateṅ or Khanār-bocan.

The frequent use of synonyms such as *khalabhūmi- khalasthala-khalasthāna* 'threshing-floor' or in the case of plant names must be explained for purely stylistic and metrical reasons. The abundance of nouns ending with the diminutive suffix *–ka* can primarily be ascribed to metrical reasons.This displays the poor poetical qualities of the compiler editor or compilers editors too.

The treatise has been composed in a metrical form and the *anuṣṭubh* is the only metre used. It is not without difficulties. Mrs. Sadhale rightly remarks 'sometimes it is felt, however, that more elaborate meter could have been used to avoid *kulakas* (clustering of several verses). One verse for one or other thought is an ideal arrangement but in KāKS a sentence or a thought-unit often spreads over two or six verses' (Se 130).

The vocabulary is rich and abounds in *hapax legomena* (cf. *Glossary*). There are some unintelligible words which remain unexplained or not satisfactorily explained for the time being such as *kaṭu* (verse 568), *kalāṭa* verse 604), *kīraka* (verse 380) , *kolakandāvali* (verse 91) *kr̥ṣṇasāri* (verse 381),*kṣudrataluka* (verse 265), *khādira* (verse 91), *camana* (verse 569), *ḍala* (verses 832, 833, 836), *pannā* (verse 785), *pecana* (verse 569), *mahāḍala* (verse 832),*rāśijaṭikā* (verses 384, 604), *lavasāraka* (verse 265), *vālaga* (verse 264) and *ṣavaka* (verses 385, 604, 620, 645, 657).

Kartana (verse 482) instead of *kaṭṭana* (Kr̥P 186), *kulastha* (verse 381) parallelry with *kulattha* (verse 521) betray hyper-correct Sanskritizms.

There are two instances of Buddhist Hybrid Sanskrit formations of past participles: *kartita* and *śuṣkita.*

Prakrit or *deśī* forms are seldom met with: *churikā* 'a short knife for cutting grain' (cf. Pāli *churikā* 'knife' PTSD 276, probably from Sanskrit **kṣurikā*; cf. *churī* = *kṣurī* ' a knife, dagger' and *chūrikā* 'knife': MW 407); *pakkakāla* (*pakka* stands for *pakva* cf. *supakva* in verse 157; the later substitutes *pakta*, cf. AirGr II 2, 866); *pādaghaṭṭanaka* (*ghaṭṭanaka* stands for **ghr̥ṣṭaṇa* cf. EWA I, 358); Kosala stands for Kośala.

Materials for the edition

Manuscripts

A This is a Devanāgarī ms. kept at the Adyar Library, Chennai (Madras), India under the signature TR - 871 – 63419 XXXVIII. I. 8. It is the transcript of a lost manuscript. According to the colophon the transcript was made by pandit V. Vijayaraghavacharya [sic!], epigraphist, Tirupati, S. India on 19th December1930. The material of the manuscript is ordinary paper. The pages are numbered. The script is clear and easily legible. There are altogether seventeen corrections made by the scribe in the text. The scribe usually put strokes over an *akṣara* or the *akṣaras* to be corrected or omitted in verses 150a, 265a, 334a, 396a, 476b, 540a, 573b, 622b, 706a, 714b and 783b. Among these there are three cases where he put the correct form in a marginal note: *matāḥ* for *guṇāḥ* (verse 293b), *saṁnirūddheṣu* for *sannirasteṣu* (verse 451b) and *kṣīrakr̥ṣṭaṁ* for *sīrakr̥ṣṭaṁ* (621a). In verses 401b and 477b he put a passage and a word respectively in parenthesis: they might have also stood in the original, but they are obviously superfluous. In the remaining cases he simply corrected his slips of pen. There is only one lacuna in the text: in verse 289b four *akṣaras* are missing. There is a metathesis in verse 729b: the manuscript reads *dhīnām* for *dhīmān*.

The manuscript contains an introductory part most likely compiled by the scribe and the text itself. The first consists of 21 numbered pages and has an index of subjects such as names of cereals, vegetables, flowers and trees as well as of certain vegetables together with their Tamil equivalents. The basic text (*mūlamātram*) comprises 113 numbered pages.

B This is a Devanāgarī ms. under the signature 38J8 belonging to Sri G. G. Joshi Pratisthan, Nagpur. It is a copy of ms. A prepared by Pandit V. N. Sastri who compared it with the original with the help of Sri N. R. Bhat. This transcript contains numerous scribal errors. At places the scribe arbitrarily constructed verses from half-verses which do not form a verse in the original but stand separately. Altogether it is much more inferior to ms. A. Moreover it is not a genuine ms. but a hastily made modern handwriting copy. I did not use it for the present edition.

Printed Editions

BU The Budapest edition was founded upon ms. A and was published by Wojtilla in *AOH* 33:2 in Budapest in 1979. The text is given Romanized transliteration where I followed the orthography of ms. A except class nasals.

Verses are consecutively numbered. I did not add any critical remark on the text. I must confess that my edition was primarily meant for drawing attention to this highly interesting text and I was fully aware of the need of a deeper study of the text which hoped to carry out after finding other mss. Unfortunately, this edition has numerous misprints which I mostly corrected later in an appendix to my English translation. In spite of these shortcomings this edition has been widely been used and referred to in studies of Indology since its publication.

SE The Secunderabad edition was published and translated by S. M. Ayachit. The Sanskrit text is based on ms. A and ms. B. Ms. B was handed over by Sri Chitale to Y. L. Nene who managed the text to get rewritten for printing. Strictly speaking this is not a printed edition but a facsimile of the handwritten text prepared under the instruction of Y. L. Nene. Ayachit got access to a photocopy of ms. A and on the basis of the latter he put his emendation of the main text in footnotes. Unfortunately, the collation of the two mss. was not always careful e.g. he simply overlooked the single lacuna in the text (289). Since the text presented in SE does not offer any significant variant reading moreover it does not reflect the southern features of ms. A, I do not use it for the preparation of my recent edition. Ayachit originally made an 'analytical rendering' which was recast by Nalini Sadhale. It is resulted in a fluent English text containing a great deal of happy solutions in translating special technical terms. However, as a whole, it is meant for other scholars than Sanskritists. To the text and translation two valuable commentaries and one appendix is added. The first commentary written by Nalini Sadhale is a careful study on the author, the contents the language and the style of the treatise. The second one is a professional interpretation from the standpoint of an agricultural expert where Y. L. Nene properly illuminates the practical side and importance of several issues propounded in the text. The appendix contains a plant index where the Sanskrit names of the plants mentioned in the work stand with their English translation, Latin botanical names and a reference to the geographical locations where they are grown in India.

Translations

Almost the whole Section One and Two were translated and published as early as 1964 by scholars in India (AAI 4–6, 20–24, 36–40 and 60–79).

Selected passages concerning the importance and management of agriculture, cultivation of rice, vegetables and garden crops, establishing gardens, accomplishing forestation and organizing markets for various goods by the king were translated by M. S. Randhawa (Randhawa 1980, 484–490).

My English translation was published in *AOH* 39:1 in 1985. My aim was to make a philologically reliable and at the same time fluent rendering of the text. I must confess that I could fulfill it only partially. A great number of terms and passages remained unsatisfactorily interpreted. The translation is accompanied with footnotes containing text critical remarks as well as cultural and professional information about certain terms and passages.

The latest English translation is referred to above in SE.

Editorial technique

The present edition is based upon ms. A, which is the single extant genuine manuscript of Kāśyapīyakṛṣisūkti. Ms B is a modern transcript without any originality moreover burdened with numerous errors therefore it has not been taken into account. The scribe was most likely a Tamil speaker therefore there are few peculiarities in orthography such as writing *gumbhitā* for *gumphitā*, *sarvadhā* for *sarvathā* (SE 130). The orthography of the ms. has a great deal of other features, too:

- conversion of consonants such as *gh* for *g* (Ghūrjara for Gūrjara);
- reduction of double to single consonants e.g. *jj* to *j* (*samujvalam* instead of *samujjvalam*) dh for ddh (*phalasidhyartham* for *phalasiddhyartham*);
- degemination of *tt* preceeding *v* (*satva* instead of *sattva, pradatvā* instead of *pradattvā*); the gemination of *t* (*ttviti* instead of *tviti* = *tu* iti);
- *ḍ* for *d* (*ḍambhāsūyādimuktāś ca* for *dambhāsūyādimuktāś ca, ḍalaṁ mahāḍalam* for *dalaṁ mahādalam*);
- *ṇṇ* for *nn* (*nirbhiṇṇādi* for *nirbhinnādi*);
- non-observance of *sandhi* rules for reasons of metre (cf. **Notes on the Text**);
- conversion of *ṛ* (*ra* for *ṛ*: *graha* for *gṛha, krantanam* for *kṛntanam*);
- use of *anusvāra* instead of the class nasal in the most cases;
- use of the class nasal instead of *anusvāra (sandṛśyate* for *saṁdṛśyate*) at places;
- omission of a non-final *akṣara* (*tadvrīkaṇikāḥ* for *tadvrīhikaṇikāḥ*);
- omission of *repha* (*vārāśijalapūreṣu* for *vārrāśijalapūreṣu*);
- omission of *visarga* (*bhuvasthale* for *bhuvaḥsthale*, *viniṣṛtam* for *viniḥsṛtam*);
- omission of *visarga* and the vertical stroke for medial *ā* (*śikṣyasvavṛttayo* for *śikṣyāḥ svavṛttayo*);
- insertion of *visarga* in final position (*yathāvidhiḥ* for *yathāvidhi*);
- omission of *anusvāra* (*prītijanaka* for *prītijakaṁ*, *savatsāś ca* for *savatsāṁś* ca);
- interchange of *bh* and *th* (*rabhadrumam* for *rathadrumam*);
- interchange of *r* for k (*bhūrjaram* for *bhūrjakam*);
- interchange of *n* for *t* (*tantuvāyakān* for *tantuvāyakāt*);
- interchange of *i* for *a* (*kriyajñānam* for *krayajñānam);*
- interchange of *e* for *ā* (^{0}*greva*^{0} for ^{0}*grāva*^{0});
- omission of *ṅ* (*ṭakayā api* for *ṭaṅkayā api*);
- addition of an *akṣara* in final position (*lakṣmīvilāsakṛtam* for *lakṣmīvilāsakṛt*);
- omission of a word (*rakṣayet tatra hi* for *rakṣayet tatra tatra hi*).

I have retained all these features in the present edition.

The text is composed in *anuṣṭubh* metre but the verses in ms. are unnumbered. Moreover the author or final editor deliberately made *kulakas* (clusters of verses). To make not too much harm to the text I simply followed the arrangement of the verses given by the scribe and only in few cases changed it for syntactical reasons. That is all is what the existence of a single ms. permits. For purely practical pusposes I have kept the numbering of my first edition with few minor changes after verse 773.

For the sake of an easy lemmatization I give the single word separately where the scribe, following the common practice in Sanskrit writing, conjoined words, whether or not they are connected by *sandhi* or compound (*samāsa*) rules. For the same purpose I gave the compounds with their members hyphenated in the Glossary; however for metrical reasons the hiatus has been kept everywhere. When the first or any subsequent member of a compound is a *hapax*, it is given as a separate entry of its own. Instead of syntactic compounds I opt for lexical compounds where each member is divided by hyphen.

Sanskrit Text

kāśyapamunikathitā

kāśyapīyakṛṣisūktiḥ

dhānyādikṛṣikramakathana-

śākādikṛṣikramakathana-

vividhahavyanivedanakramakathanātmaka-

bhāgacatuṣkaguṁbhitā

(mūlamātram)

1. tatra śāstropadeśakramaḥ

kāśyapa upadiśati

dharmavit kṣatriyo rājā prajārakṣādhuraṁdharaḥ |
daṇḍayitvātha daṇḍyāṁs tu satyaṁ dharmaṁ ca sarvataḥ ||1||
sthāpayitvā viśeṣeṇa varṇāśramaratān api |
prajāḥ suprakṛtiḥ kāle poṣayet devatuṣṭaye |
devāś ca yakṣagandharvāḥ pitaro munayaś ca ye ||2||
ye cānye makhahomādikriyārhāḥ saṁprakīrtitāḥ |
tatprīṇanaṁ tu yajanaiḥ devayajñaṁ vidur budhāḥ ||3||
prajānām atithīnāṁ ca prīṇanaṁ bhojanādibhiḥ |
dhenvādiprīṇanaṁ kālyaṁ bhūtayajñaṁ prakīrtitam ||4||
ṛṣiyajñādayo ye ca nirdiṣṭā munibhiḥ kramāt |
vedavidyādānajapādānais tatprīṇanaṁ matam ||5||
ato yajñeṣu sarveṣu devayajñaḥ pradhānakaḥ |
dvitīyo bhūtayajñas tu tannidānaṁ mataṁ mama ||6||
tasmāt sarveṣu deśeṣu bhūmiśo rakṣaṇavratī |
devayajñaṁ bhūtayajñaṁ dvayam eva prapālayet ||7||
śuddhadravyair devayajñaḥ saphalaḥ kathito budhaiḥ |
tadarhadravyam ādyaṁ tu sasyādiḥ parikīrtitaḥ ||8||

sasyodayaphalā bhūmiḥ dhṛtā vārāharūpiṇā |
viṣṇunā bhūṣṇunā pūrvaṁ pātālajalapātataḥ ||9||
saṁprārthitas tayā devyā devaḥ śārṅgiparaḥ pumān |
harṣeṇa bhūmyā māhātmyam avadat surasannidhau ||10||
ratnādidhāraṇāt garbhe ratnagarbhā vasundharā |
dhānyādīnām oṣadhīnāṁ jalānāṁ prasravād api ||11||
dharā tu medinī khyātā prāṇināṁ prāṇavardhinī |
dvipadā ye catuṣpadāś cāṇḍajā vividhās tu ye ||12||
te sarve prāṇivargās tu medinyāḥ sāravaibhavāt |
prāptasatvāḥ prāṇadharāḥ jīvino niścitā bhuvi ||13||
pannagānāṁ tathāny eṣāṁ dvipadāṁ ca catuṣpadām |
bharaṇāt prāṇadānāc ca devaprītiś ca śāśvatī ||14||
satyadharmācaraṇakaṁ caturvargaphalātmakam |
tataḥ sṛtis tu saphalā tena modati sṛṣṭikṛt ||15||
ataḥ sasyaṁ dharāprāṇaḥ dharāyā vṛttir uttamā |
māṅgalyasūtraṁ ca tathā sasyam āhur divaukasaḥ ||16||
sasyādir eva medinyāḥ jīvanāḍī kalātmikā |
sasyādir eva medinyāḥ paro dharmaḥ paraṁ yaśaḥ ||17||
sasyapūrṇā vasumatī prāṇīnāṁ prāṇavardhinī |
sarvamaṁgaladhātrī ca devānāṁ tuṣṭidāyinī ||18||
nityasasyā ca madhurajalasrāvā viśeṣataḥ |
śobhate sarvataḥ pṛthvī viṣṇor ājñāvalaṁbinī ||19||
viṣṇunā pālitā bhūmiḥ jananī prāṇasantateḥ |
poṣayitrī cādyamātā puṣṭidā tuṣṭīdā tathā ||20||
ratnagarbhā paraṁ nityaṁ sarvarakṣaṇakāriṇī |
parvatottuṁgavakṣojā payodhiraśanā tathā ||21||
samudramuktā bhūṣāḍhyā devī rājati śārṁginaḥ |
sūryacandrāgnidīpeyaṁ mandamārutavījanā ||22||
viśālavanasaṁcārā pṛthivī viṣṇuvallabhā |

sarvaṁsahā rājate 'tra dikcakraracitasthitiḥ ||23||

tām etāṁ vasudhāṁ kṛtsnāṁ kṛṣiyogyasthalojvalām |

nānādeśeṣu vibhajet nānākārāṁ mahīpatiḥ ||24||

(2) girinadīnadavanasārakṣetrādibhedena

sasyārhabhūmivibhajanakramakathanam

kāśyapa upadiśati

payodhisaṁvṛtā bhūmir iyam āsṛṣṭi sarvataḥ |

brahmaṇā tu vidhātrā sa nirmitā hi samasthalā ||25||

phalapradātrī vividhā kalpeṣu vividheṣv api |

ślāghitā devagandharvamunibhūpālavṛndakaiḥ ||26||

prāptasatvaphalā bhūmiḥ pratikalpaṁ ca bhāvitā |

kvacin nimagnā vārāśijalapūreṣu bhūriyam ||27||

kvacit tathā samudreṇa tyaktas vāntapradeśakā |

kvacin nirbhiṇṇanāḍī ca kālabhedena kalpataḥ |28 ||

sārāsārasravayutā nimnā connatarūpiṇī |

giribhiḥ saṁvibhaktā ca nadībhiś ca kvacit sthale |

nadena sārabhūmyā ca hradena mahatāpi ca ||29||

kvacit śarkararūpā ca kvacid atyuṣṇarūpiṇī |

kvacit jalavihīnā ca jāteyaṁ vasudhā kramāt ||30||

kvacid ūṣararūpā ca kvacit bījavināśinī |

evaṁ bahuvidhā jātā medinīyaṁ prakīrtitā |31||

ato mahīpatiḥ sarvaprāṇipālanadīkṣitaḥ |

sasyābhivṛddhaye bhūmiṁ parīkṣeta sulakṣaṇām ||32||

asthipāṣāṇakhaṇḍādyair hīnāṁ mṛdulamṛttikām |

susnigdhām alparaktāṁ ca kṛṣṇavarṇāṁ tathaiva ca ||33||

tuṣakācavihīnāṁ ca sārāṁ rasasamujvalām |

na śvabhrayuktāṁ nāgādhāṁ nottuṁgāṁ ca samasthalām ||34||

mallikājātikuṭajasurāgandhasamujvalām |

atha vā padmakharjūratiniśaprasavakramām ||35||

apītasalilāṁ śaśvat pītodakajalām api |
bījavṛddhikarīṁ vegāt sītāsaukhyapradāyinīm ||36||
vṛṣaphenāktakāṁ vāpi satvajantusamanvitām |
akaṁṭakāṁ karīṣādihīnāṁ ca nibiḍāṁ ghanām ||37||
tolane 'dhikabhāraḍhyām oṣadhivṛddhidām api |
brāhmaṇakṣmāṁ kṣtrabhūmiṁ vaiśyabhūmiṁ tathaiva ca ||38||
śūdrabhūmiṁ kvacit deśe saṁkīrṇaguṇabhūmikām |
evaṁ pañcavidyāṁ jātyā vibhaktāṁ vimalodakām ||39||
antaḥ sārayutāṁ bāhyasārām api ca kalpakīm |
nayanānandajananīṁ valmīkādivivarjitām ||40||
duṣṭasatvavihīnāṁ ca satvapakṣiniveśinīm |
vātavegena vātyā vā vahnivegena vā punaḥ ||41||
apītasalilalām antarniṣkuṭodyānavṛddhidām |
pracchāyabhūrūhāṁ vṛddhidāyinīṁ vālpanimnakām ||42||
nānābījāvalivṛddhikārinīṁ ca muhurmuhuḥ |
mṛdusparśāṁ ca nitarāṁ gopriyāṁ godhanapriyām ||43||
śubhalakṣaṇasaṁyuktām alpadoṣām api kvacit |
sulabhodakanisrāvāṁ sulabhasvīkṛtodakām ||44||
evaṁ lakṣaṇasaṁyuktakhalabhūmivṛtāṁ kvacit |
vasudhāṁ bhūpatir vīkṣya grahnīyād uttamām iha ||45||
bhūparīkṣākramavido guṇāḍhyā nṛpacoditāḥ |
brāhmaṇāḥ kṣatriyā vāpi viśaḥ śūdrās tu vā punaḥ ||46||
dagārgalapramāṇajñāḥ kṛṣiśāstraviśāradāḥ
parīkṣayeyuḥ bhūmiṁ tu śubhe śubhanimittake ||47||
snātaḥ śuciḥ śuddhavāsāḥ dhṛtasvarṇāṁgulīyakaḥ |
nirdiṣṭāṁ bhūmim āsādya kuryāt tatra pradakṣiṇam ||48||
saha maṁgalaghoṣeṇa svastivākyena kovidaḥ |
sāyaṁkāle 'tha vā prātaḥ brāhmaṇair vedapāragaiḥ ||49||
paṁcagavyena śuddhena salilenātha vā punaḥ |

khanitreṇāpi tulayā ṭakayāpi kvacit budhaḥ ||50||
bhūmiṁ parīkṣayet khātvā nirīkṣya ca muhurmuhuḥ |
āghrāya pītvā ca mṛdaṁ tolayitvāpi vā kvacit |
kṣiptvā pātrasthasalile ghaṭikāvadhi kovidaḥ ||51||
varṇabindurasasrāvaiḥ vidyāt bhūmyāś ca lakṣaṇam |
samavarṇā samacchāyā ghanā snigdhā ca bhūr api ||52||
devānām api bhūpānāṁ praśastā muniniścitā |
brāhmaṇānāṁ viśāṁ cāpi kṣemadā vṛddhidāyinī ||53||
śubhalakṣaṇasaṁyuktā sarveṣāṁ śubhadā dharā |
kuṭumbārogyadā śaśvat dhanagodhānyavṛddhidā ||54||
nirdiśed uttamāṁ bhūmiṁ madhyamām api vā kvacit
adhamāṁ varjayet yatnāt bhūmilakṣaṇapaṇḍitaḥ |
kṛṣiyogyā bhūmir iyam udyānāyeyam āhatā ||55||
vanāya bhūriyaṁ proktā jalādhārāya niścitā |
iyaṁ bhūmir iti proktā caturthī ca svabhāvataḥ ||56||
iti bhūmiṁ śubhe kāle nirdiśec ca nṛpājñayā |
sīmāsthalavibhāgaś ca kartavyo bhūmim icchatā ||57||
grāmeṣv api ca deśeṣu durgeṣu nagareṣv api |
prāsādāntas tu vā bhūmiṁ kṛṣyudyānādiyogyakām ||58||
nirdiśen matimān yuktyā nānākṣetrasvarūpavit |
jalasvarūpavit prājñaḥ prāṇināṁ sukhasiddhaye ||59||

(3) bhūmyāṁ jalādhārasthāpanakramakathanam

kāśyapa upadiśati

pramāṇajñaiḥ śilpivaryaiḥ śāstrapāraṁgatair nṛpaḥ |
kārayet grāmanirmāṇaṁ durgāṇāṁ ca purām api ||60||
vanamadhye nadītīre girisānutaṭe 'tha vā |
svadeśeṣv api rāṣṭreṣu tatra tatra viśeṣataḥ ||61||
ābhīrapalliṁ ghoṣaṁ ca vrajaṁ vā godhanāśritam |
tathā kṣudragrāmajālaṁ sthāpayet sumuhūrtake ||62||

vāpīkūpasamāyuktam udyānādivibhūṣitam |
sasyakṣetrāvṛtaṁ kāryaṁ niyamena mahīpatiḥ ||63||
grāmāṇāṁ nagarāṇāṁ ca purāṇām api sārabhūḥ |
sasyārhā kathitā prāṇaḥ jīvanaṁ cottamaṁ tathā ||64||
prāṇanāḍīti gārgyaś ca provāca vadatāṁ varaḥ |
sasyārhakṣetrasaṁyogād eva sarvatra śāśvatam ||65||
grāmādinirmāṇaphalaṁ kathitaṁ munipuṁgavaiḥ |
prāṇirakṣā ca nitarāṁ puruṣārthaphalaṁ tathā ||66||
devatuṣṭiyajñaphalaṁ sṛṣṭeś ca phalam uttamam |
sajīvasya hi lokasya prāṇarakṣākarī bhṛśam ||67||
sārabhūr iti nirdiṣṭā jīvānāṁ phalam uttamam |
ataḥ sulakṣaṇopetaṁ doṣahīnāṁ mahīpatiḥ ||68||
atha vā svalpadoṣāṁ tu nirdiśet tatra tatra tu |
vibhaktāṁ grāmamārgeṇa puramārgeṇa vā kvacit ||69||
mahāmārgeṇa vā deśe vibhajet bhūmim uttamām |
grāmāṇāṁ nagarāṇāṁ ca paścime bhāga uttame ||70||
udīccyām atha vā prācyāṁ dakṣiṇasyāṁ tathā kvacit |
jalāśayaṁ sthāpayet tu bhūmisthityanusārataḥ ||71||
gabhīraṁ parighāvat tu dhanurākārabhāsuram |
dīrghaṁ kvacid vartulaṁ vāgādhaṁ tu salilāśayam ||72||
dṛḍhaprapālikasetumārgeṇa ca surakṣitam |
uttuṁgataṭakalpāḍhyam atho mārgeṇa saṁyutam ||73||
jalāgamanamārgeṇa śobhitaṁ ca viśeṣataḥ |
ato gireḥ samīpe vā cottuṁge bhūtale 'tha vā ||74||
mahāhradena saṁyukte sthāpayet tu jalāśayam |
mahājalasthairyabhājaṁ jalādhāraṁ tu bhūtale ||75||
sārakṣetre sthāpayet tu svayaṁ srāvayutepi vā |
saikatādivihīne tu pāṣāṇādivivarjite ||76||
jalāśayaṁ tu gaṁbhīraṁ jalapūrārgalāyutam |

tathā taṁ sthāpayet vidvān jalapūrṇe jalāśaye ||77||

yena vā na virodhaḥ syāt grāmādeḥ nagarasya ca |

prākārodyānadurgāṇāṁ khalakṣetrasya vā punaḥ ||78||

jalabhītiḥ pravāhādyaiḥ na syāt sarvatra bhūmipaḥ |

tathā taṁ sthāpayet deśe parīkṣeta kramād imam ||79||

māsakrameṇa vā varṣakrameṇa ca muhurmuhuḥ |

sthāpayen nṛpatis tatra taṭarakṣaṇakāriṇaḥ ||80||

varṣākāle viśeṣena kulyādīnāṁ śataṁ tathā |

vāpīnāṁ dīrghikāṇāṁ ca rakṣaṇaṁ kṣemadaṁ matam ||81||

jalāśaye tu salilapūrṇe saṁrakṣite kramāt |

sasyādivṛddhis tatraiva niścitaḥ sukhadāyakaḥ ||82||

rogahānir vahnibhayahāniś ca sukham uttamam |

poṣaṇaṁ rakṣaṇaṁ śastaṁ dvipadāṁ ca catuṣpadām ||83||

vihaṁgamānām api ca sarpādīnāṁ ca sarvataḥ |

sukhaṁ tu paramaṁ coktaṁ na vyādhiś ca bhavet dhruvam ||84||

pānadravyaṁ tu salilaṁ dvipadāṁ ca catuṣpadām |

ekam eva hi nirdiṣṭaṁ brahmaṇā sṛjatā prajāḥ ||85||

mahodakasthānam eṣaḥ jalāśaya udīritaḥ |

ataḥ sarvaprayatnena prajārakṣādhuraṁdharaḥ ||86

sthāpayet rakṣayed rājā jalāśayam imaṁ kramāt |

jalāśayadvayaṁ bhūpaḥ kvacid ekaṁ jalāśayam ||87||

sārakṣetrānusaraṇaṁ sthāpayet grāmayogyakam |

giriprasravaṇopetaṁ mahāhradayutaṁ tathā ||88||

vananimnagayā yuktāṁ mahānimnagayā kvacit |

deśe deśe viśeṣeṇa grāmāṇāṁ ca purām api ||89||

jalavyayasthānayutaṁ kalpayet yuktitaḥ kramāt |

jalāśayataṭe bhūpaḥ vṛkṣāṁś ca parivardhayet ||90||

vetasaṁ ca vaṭaṁ tatra kolakandāvaliṁ kvacit |

plakṣam aśvatthakaṁ caiva khadiraṁ khādiraṁ tathā ||91||

tindukaṁ tiṁtriṇīṁ vāpi bhūrjaraṁ cāpi veṇukam I
niṁbaṁ kadaṁbam api ca ye ca sāradrumā matāḥ II92II
tāṁs tu vṛkṣān mahīpālaḥ yatnenāhṛtya kālataḥ I
jalāśayataṭe yuktyā sthāpayet kṣemasiddhaye II93II
taṭasya nimnabhūmyāṁ vā tatkulyāyāś ca nirgame I
snānaghaṭṭāvirodhena pānaghaṭṭārhakaṁ kvacit II94II
vṛkṣāṁś ca vardhayet yuktyā tatra tatra viśeṣataḥ I
khalabhūmyāṁ tathā vṛkṣān vaṭādīn sukhadāyinaḥ II95II
taṭakadīrghikātīre hradādīnāṁ taṭe tathā I
pracchāyavṛkṣān bhūpālaḥ vardhayet kramatas tathā II96II
ākrīdeṣv api codyāneṣv evaṁ viśrāntibhūmike I
mahāmārgasthale tatra jalāśayasamīpataḥ II97II
vātapotaprasavakān prāṇināṁ śramahāriṇaḥ I
vanabhojanayogyeṣu sthāneṣu vividhesv api II98II
devālayodyānatale bhūpālodyānake tathā I
munīnām uṭajānāṁ ca tale tatra viśeṣataḥ II99II
senāniveśasthāne ca bhūpaviśrāntikasthale I
yajñārhasthalake cāpi balyarhasthānakepi ca II100II
pracchāyavṛkṣān bhūpālaḥ vardhayet kramaśas tathā I
vihaṁgamāhlādakarān phalapuṣpādidāyinaḥ II101II
tathā vyādhipraśamakān śītalān komalān tarūn I
akaṁṭakān ca panasān āmrāṁś cāpi rasālakān II102II
punnāgamālatīkundacampakāvalim apy atha I
sthāpayed āvaraṇake tatra tatra viśeṣataḥ II103II
jalāśayajalasrāvāt vardhayec ca kramāt tarūn I
jalāśayavihīne tu sukhaṁ naivopajāyate II104II
na sāndhyakarma na snānaṁ vṛkṣāṇām api vardhanam I
na syād ity eva munibhiḥ saukhyaṁ ttv iti viniścitam II105II
ataḥ sarvaprayatnena bhūpālo rakṣaṇavratī I

grāmeṣv api ca deśeṣu vanamadhye tu vā punaḥ ||106||

sthāpayitvā tu gaṁbhīraṁ jalāśayam ataṁdritaḥ |

kulyāṁ dṛḍhataṭāṁ nityaṁ sulabhodakanisravām ||107||

nānāśākhāsamākrāntāṁ mahākulyāyutāṁ tathā |

jalāśayena sahitāṁ sthāpayet bahuśo dṛḍhām ||108||

(4) atha nadīmātṛkakulyākūpādikramakathanam

kāśyapa upadiśati

yatra sasyakṣiteḥ kulyā jalādhāropi vā kvacit |

na kalpitā vā na sthāpyā satatodakapūritā ||109||

tatra kulyā nadīmātṛmukhā sthāpyeti niścitā |

tat caturvidham ākhātaṁ kṛṣipaddhatikovidaiḥ ||110||

ādyā nadīmātṛkā syāt kulyā cātha hradāśritā |

jalāśayaśritā vāpi mahīdhrabhṛgupārśvabhāk ||111||

kulyā caturvidhā hy evaṁ sthāpyā deśapramāṇataḥ |

yatrordhvabhāge nadyādiḥ dṛśyate bhūsvabhāvataḥ ||112||

tatra nadyā mukhaṁ kāryaṁ svataḥ prasrāvabhāsuram |

kṛṣiyogyaṁ sthalaṁ yatra vidyate deśabhūmiṣu ||113||

prāyas tadūrdhvake bhāge dūre vā dūratopi vā |

kulyādhāraḥ prakartavyaḥ kṛṣikāryavicakṣaṇaiḥ ||114||

prākgāminyāṁ nimnagāyāṁ kulyāraṁbhaṁ kvacit sthale |

tathā paścimagāminyāṁ kulyāṁ saṁparikalpayet ||115||

haridantaragāminyām atha vā tāṁ prakalpayet |

koṇāntarapraveśinyām āpagāyām api kvacit |

krameṇottuṁgabhūmiṣu sarvatra hi viśiṣyate ||116||

nimnagā mātṛkākulyā nadyārabdhā kṛtasthalā |

krameṇa bhūmiṁ vividhām ākramya bahurūpikām ||117||

grāmāt grāmam athākramya vanād vanatalaṁ tathā |

svakīyakṣetragāṁ kuryāt prakṛtikṣemasiddhaye ||118||

caturhastaḥ paṁcahastaḥ ṣaṭhastaḥ saptahastakaḥ |

daśahastakamānaṁ vā kulyāvaipulyam īritam ||119||
kvacin mahāhradārabdhā kulyā sā jalapūritā |
tathā mahodakanadītalārabdhā ca niścitā ||120||
svakṣetrabhāgāvadhikaṁ kulyāvaiśālyam uttamam |
tulyam evaṁ hi kalpyaṁ tad unnatis tu sthalocitā ||121||
sarvatrāgādhatā yojyā jalaprasrāvayogyakā |
mahāṭatākaṁ tīrtvā vā madhyamārgaṁ kvacit tale ||122||
sarvaprāṇihitāyaitāṁ kulyāṁ bhūpaḥ prakalpayet |
giriprasravaṇārabdhāṁ kulyām api tale kvacit ||123||
svakṣetragāminīṁ kuryāt kṛṣisaṁpadvivṛddhaye |
kulyām ekāṁ pūrṇajalāṁ tathā kulyādvayaṁ kvacit ||124||
kulyātrayaṁ vā kṣetrāṇāṁ hitāya hi vidhīyate |
nadīhīne kvacit deśe saikatasthalakalpitā ||125||
kulyāpi ca sukhāyaiṣā niścitā kṛṣikovidaiḥ |
santatasrāvasaṁyuktā sthalī sā grāvavarjitā ||126||
sukṣetrā suguṇopetā sasyakṣemāya kalpate |
hradārabdhāṁ tu vā kulyāṁ neyāt svagrāmasaṁpade ||127||
udyānāyāpi vā deśe cākrīḍāyāpi vā kvacit |
kulyāprakalpanaṁ yuktyā jalasrāvasamujvalam ||128||
snānāya pānāya tathā kṛṣikāryāya ca kṣitau |
niścitaṁ munibhiḥ pūrvavedibhiḥ jīvarakṣakam ||129||
kvacit deśe bhūtalaṁ tu kulānīraṁ praṇāśayet |
pāyayet śoṣayed vāpi svāntardoṣādibhiḥ svayam ||130||
atas tat bhūtalaṁ varjyaṁ kulyānirmāṇakovidaḥ |
kvacid ūṣarabhūmiś ca dṛśyate sarvataḥ kṣitau ||131||
sā cāpi varjyā kulyāyai bhūdoṣākrāntabhūr api |
ataḥ satguṇasaṁpannaṁ bhūtalaṁ vīkṣya pārthivaḥ ||132||
kulyāprakalpanaṁ deśe sthāpayet sasyasaṁpade |
kvacit kulyādvayaṁ yojyaṁ tathā kulyātrayaṁ kvacit ||133||

sthalānukūlaṁ sarvatra tatkalpanam udīritam
sarvāsām api kulyānāṁ jalādharasthalīṣu ca ||134||
praveśaś ca kramāt proktaḥ śāstravidbhiḥ munīśvaraiḥ |
jalāśayavihīne tu sthale grāme pure tathā ||135||
vane vanāntare vāpi tāṁ kulyāṁ kṣetragāminīm |
prakalpayet tathā dhīmān sasyavṛddhyai viśeṣataḥ ||136||
tale tatra kṣudrakulyā śasyate ca phalapradā |
prāṇinām api sasyānāṁ jīvanaṁ jīvanaṁ matam ||137||
ataḥ sarvatra deśeṣu pārthivaḥ kṣemasiddhaye |
kulyāṁ nadīmātṛkāṁ vā sthalāntarakṛtām api ||138||
jalapūrṇāṁ prakurvīta khātayet rakṣayet sadā |
arakṣitā sā viphalā doṣabhāk jalavarjitā ||139||
ato bhūpatibhiḥ sarvaiḥ kulyārakṣaṇam uttamam |
dharma ity eva nirdiṣṭam ṛṣibhis tattvadarśibhiḥ ||140||
vāpībhyaś ca taṭākebhyaḥ kūpebhyaś ca viśeṣataḥ |
jalapoṣaṇam ādiṣṭaṁ kṛṣikṣemāya bhūtale ||141||
kvacit sthalyām ūrdhvabhāge niścitā salilasthitiḥ |
kvacit tu kathitā bhūmyām adhobhāge jalasthitiḥ ||142||
kvacit pāṣaṇabhūyiṣṭhe jalasrāvas tu naśyati |
mṛddoṣadūṣito 'nyatra jalasrāvaś ca naśyati ||143||
ataḥ saikatabhūyiṣṭhā jalapūrṇā vasundharā |
vāpikūpādiyogyeti niścitā tattvadarśibhiḥ |144||
yasmin deśe grīṣmakāle sasyakṣetrāya ca punaḥ |
udyānāyāpi vā nīraṁ na paryāptaṁ hi kulyajam |145||
tasmin tu deśe sarvatra kūpaṁ nirmāpayen nṛpaḥ |
kṣudrakūpaṁ ca kūpaṁ ca mahākūpaṁ ca vāpikām ||146||
caturaśraṁ maṇḍalaṁ vā dīrghākāram athāpi vā |
sthāpayet khātayet tac ca rakṣayec ca yathāvidhi ||147||
uttarāyanamāseṣu prāyo deśeṣu sarvataḥ |

adhobhāge jalaṁ dṛśyaṁ nadyām api tale kvacit ||148||
tasmāt kūpādikhananam uttarāyaṇamāsike |
saṁpūrṇasalilāvāptihetave cirakālikam ||149||
dagārgalavidhijñena dhīmatā bhūmivallabhaḥ |
vāpikūpādikhananaṁ salilasthitim eva ca ||150||
ādau niścitya vṛkṣādyaiḥ bhūnāḍīvīkṣaṇād api |
bhūmiṁ parīkṣya kalayet sumuhūrte sulagnake ||151||
prāyaḥ prātas tu khananaṁ śubhāya parikīrtyate |
abhyarcya varuṇaṁ nīranāthaṁ bhūdevikām api ||152||
vanadevīṁ tathābhyarcya balidānāt viśeṣataḥ |
kūpaṁ vāpīṁ dīrghikāṁ ca sthāpayet kramato nṛpaḥ ||153 ||
ādau mahāvaṭaṁ khātvā dīrghaṁ vā caturaśrakam |
maṇḍalākāram atha vā khātamṛtsnāṁ tu dūrataḥ ||154||
nikṣipya kramaśas tatra jalasandarśanāvadhi |
mṛduccharaṇam ākhyātaṁ bhaṭaiḥ vīraiḥ kṛṣīvalaiḥ ||155||
jalaṁ saikatasaṁmiśraṁ vīkṣya kūpādiṣu kramāt |
adho 'dhiṣṭhānakalpas tu dārḍhyāya ca vidhīyate ||156||
sthale saikatabhūyiṣṭhe ceṣṭikājālakair api |
supakvair iha cādiṣṭam adhiṣṭhānaprakalpanam ||157||
kaṭhine bhūmibhāge tu dīrghikākalpena kvacit |
adhiṣṭhānam adhobhāge śilākhaṇḍaiḥ prakalpayet ||158||
kūpasyādhaḥ sthale caivam adhiṣṭhānaṁ yathākramam |
sikatājalasaṁmiśramṛduccharaṇakāryataḥ ||159||
dṛḍhīkṛtya kramān nityaṁ bhūyiṣṭhaṁ salilaṁ tataḥ |
vilokya tatra kūpādāvadhiṣṭhānopari kramāt ||160||
iṣṭikākhaṇḍakaiḥ kāle nirmāṇaṁ kalpayen nṛpaḥ |
sudhāsaṁmiśritaiḥ kūpanirmāṇaṁ ttv iṣṭikādibhiḥ ||161|
cirakālasthitikaram āhus tattvavido budhāḥ |
ataḥ sarvatra deśeṣu vāpīkūpādikalpanam ||162||

sudheṣṭikādibhiḥ kāryam iti śāstreṣu niścitam |
adhiṣṭhānāt bhūtalāntaṁ pratyahaṁ seṣṭikādibhiḥ ||163||
prakalpanaṁ prakurvīta kvacit sopānakalpanam |
mukhadvāraṁ prakurvīta śilābhiḥ bhūtale kramāt ||164||
prācyāṁ prāticyām atha vā sthalayogyaṁ vidur budhāḥ |
ghaṭiyantrasthalaṁ tīre śilābhiḥ parikalpayet ||165||
tataḥsalilanisrāvahetave dṛḍhabhūmike |
tīreṣu kṣudrakulyāṁ ca sthāpayet sthalayogyakam ||166||
ghaṭīyantraṁ tu vividhaṁ vṛṣabhair vāhyam uttamam |
dṛḍhaśṛṁkhalikāyogāt hastihastādibhiḥ kvacit ||167||
vāhyaṁ tu tan madhyamaṁ syād adhamaṁ naravāhyakam |
ghaṭīyantrapreraṇena cādhaḥsthaṁ salilaṁ kramāt ||168||
kūpāditīrabhāk loke dṛṣṭaṁ sarvatra sarvataḥ |
tan nīraṁ kulyakāyogāt sasyakṣetrāya śasyate ||169||
kṛṣikāryaratas tasmāt dhīmantaḥ puruṣā bhuvi |
brāhmaṇāḥ kṣatriyā vaiśyāḥ śūdrāś cānyepi yuktitaḥ ||170||
ghaṭīyantraprayogena vāpīkūpādinīrakam |
jalāśayasthaṁ nīraṁ vā nadīmātṛkakulyagam ||171||
hradotthaṁ ca jalaṁ śuddhaṁ jaladoṣavivarjitam |
kṛṣikṣetrāya śākādisthānakāya ca bhūtale ||172||
caṁpakādilatāsthalyām api nītvā yathākramam |
kṛṣīprayojanaṁ loke prāpnuvanti yathāvidhi ||173||
jalādhārasthalaṁ tasmād ādau prāpyaṁ kṛṣīvalaiḥ |
kulyādyair api kūpādyaiḥ hradādyair api bhūtale ||174||
jalāśayādyair api ca jalaṁ prāpyaṁ hi niścitam |
meghapracārakāle tu vṛṣṭiḥ sarvatra niścitā ||175||
daivayogān munīndrāṇāṁ saṁkalpād api sarvadhā |
kādaṁbinībhiḥ kāle tu saṁvṛṣṭaṁ salilam nṛpaḥ ||176||
jalāśayādisthāneṣu pūrayet kṣemasiddhaye |

rakṣayet tatprayatnena jalamūlā kṛṣir matā ||177||
ataḥ sarvatra bhūpālair anyaiḥ puruṣapuṁgavaiḥ |
ṛtukāle tu sarvatra jalaṁ prāpyaṁ prayatnataḥ ||178||
saṁrakṣyam iti ca proktaṁ kāśyapena maharṣiṇā ||179||

(5) kṛṣīvalānāṁ grāmakāryakarāṇāṁ ca guṇalakṣaṇakathanam | kṛṣīpraśaṁsā ca kāśyapa upadiśati

kṛṣikāryaratā ye tu puruṣāḥ grāmavāsinaḥ |
ḍaṁbhāsūyādimuktāś ca parasparahitaiṣinaḥ ||180||
gobhūmīdevabhaktāś ca nitarāṁ satyavādinaḥ |
parānukūlaniratāḥ santataṁ tuṣṭacetasaḥ ||181||
ajātaśatravo loke parakāryaprayojanāḥ |
catuṣpādvargavātsalyabhūṣitā nayakovidāḥ ||182||
gaṇanākarmakuśalāḥ śuddhāḥ śaucaparāyaṇāḥ |
nityakarmaratāḥ kāle kṛtātithisusatkriyāḥ ||183||
tandrālasyādihīnāś ca kāmakrodhādivarjitāḥ |
parasparaṁ snehabhājaḥ sāhyakarmaratāś ca ye ||184||
te tūttamāḥ samādiṣṭāḥ puruṣāḥ puṇyadarśanāḥ |
jalāśayataṭādīnāṁ kulyādīnāṁ ca rakṣakāḥ ||185||
gorakṣaṇaikaniratāḥ rājaviśvāsaśālinaḥ |
putrapautrādisaṁpannāḥ śvaśurabhrātṛvargakāḥ ||186||
brāhmaṇāḥ kṣatriyā vaiśyāḥ śūdrāś ca kṛṣijīvinaḥ |
saṁkīrṇavarṇajā ye tu vyādhādyā ye narā bhuvi ||187||
ākheṭarasikā ye tu ye ca gorakṣakā matāḥ |
ye tv ajāpālakāḥ khyātāḥ ye cānye kṛṣijīvinaḥ ||188||
te sarve manujā loke sarvadeśeṣu sarvadhā |
parasparaprītiyuktāḥ lokakṣemāya dīkṣitāḥ ||189||
grāmāgragaṇyaiḥ jātyagragaṇyair evaṁ nṛpājñayā |
śikṣyasvavṛttayo lokavārtābhītāḥ kṛpālavaḥ ||190||

svasvajīvanarakṣākhyamakheṣu kṛtadīkṣakāḥ |
sarvaprāṇihitāyaite kathitāḥ puṇyapūruṣāḥ ||191||
teṣu ye brāhmaṇāḥ śuddhāḥ satyadharmaparāyaṇāḥ |
kṛpālavo nītividaḥ paścāt tāpādibhūṣaṇāḥ ||192||
yathārhadaṇḍakāḥ sarvarakṣākarmaṇi dīkṣitāḥ |
nṛpālājñāpālanotkāḥ bahusadguṇamaṇḍitāḥ ||193||
grāmādhipatayas te tu kartavyā bhūtim icchatā |
grāmasyaikasya vā te tu grāmāṇāṁ vā kvacit sthale ||194||
nyāyyānyāyyakramasthairyadharmasthāpanakāriṇaḥ |
te vai kṛṣikriyāsāhyakāriṇaḥ saṁprakīrtitāḥ ||195||
bhūpālaiḥ kṣatriyair evaṁ dhanikair vaiśyakair api |
kṛṣikāryeṣu taiḥ sāhyaṁ kāryaṁ lokahitāya hi ||196||
dharmyaṁ yaśasyam āyuṣyaṁ kṛṣisāhyam itīritam |
bhūpālaiḥ kṣatriyais tasmāt brāhmaṇaiḥ vaiśyakair api ||197||
śūdrair api yathāśakti kṛṣisāhyaṁ mahāphalam |
nānāvidhānāṁ bījānāṁ gavām api ca dānataḥ ||198||
kṣetradānān mahāpuṇyam āhur loke munīśvarāḥ |
kulyākhananato vāpi sasyakṣetrasya rakṣanāt ||199||
jalāśayasthāpanād vā vāpyādīnāṁ viśeṣataḥ |
annaśālāsthāpanād vā kṛṣisāhyaṁ mahāphalam ||200||
gavāṁ ca saṁrakṣaṇato brāhmaṇānāṁ viśeṣataḥ |
yathāśaktikriyāsāhyaṁ mahāphalam udīritam ||201||
tasmāt sarvatra deśeṣu puruṣās tu kṛpālavaḥ |
grāmarakṣaṇabhṛtyāṁś ca taṭarakṣaṇabhṛtyakān ||202||
sthāpayitvā hi sarvatra tatra tatra viśeṣataḥ |
vidrāvya caurān daṇḍyāṁś ca daṇḍayitvā yathākramam ||203||
dharmasthitiṁ ca paramaṁ sukhaṁ jīvanarakṣaṇam |
prakurvantas tu puruṣāḥ devaiḥ ślāghyāḥ prakīrtitāḥ |
satyaṁ dānaṁ kṣamā śīlam ānṛśasyaṁ tapo ghṛṇā ||204||

yasmin sandṛśyate sarvaṁ niyoktavyaḥ sa puruṣaḥ |
grāmanyāyasabhāyāṁ ca bhūmipālair yathākramam ||205||
grāmādhipatyaṁ ca tathā kṛṣiyuktipravartanam |
tadadhīnaṁ śubhakaram ity āhuḥ paramarṣayaḥ ||206||
brāhmaṇānām alābhe tu kṣatriyo vaiśya eva vā |
tatra saṁyojanīyaḥ syāt prajānāṁ kṣemahetave ||207||
sarvatrārṣaṁ pramāṇaṁ ca kṣemāya vijayāya ca |
kṛṣiyuktis tu loke 'smin kvacit śūdreṣv api sthitā ||208||
sa śūdraḥ sadguṇāḍhyaś ca viprabhaktiparāyanaḥ |
śāstraviśvāsaśālī ca mānajñaḥ sādhusevakaḥ ||209||
dhīmān kālapramāṇajñaḥ hitakārī hitapriyaḥ |
bhavitavyaś ca vinayī tadyuktiḥ ślāghanāspadam ||210||
prāyo grāmeṣu sarvatra bhṛtyāḥ śūdrāḥ prakīrtitāḥ ||211||
ta eva kṛṣikāryeṣu yoktavyā nānyajātayaḥ |
vaiśyāś ca kṣatriyāś caiva guṇāḍhyāḥ sūkṣmabuddhayaḥ ||212||
gaṇanādikriyāsv evaṁ rakṣaṇādikriyāsu ca |
niyoktavyā bhūmipālaiḥ viprā nītiṣu nityaśah ||213||
gavāṁ rakṣaṇakāryeṣu prajārakṣaṇakarmani |
corādidaṇḍakāryeṣu jalādhārādirakṣaṇe ||214||
devalāyādirakṣāsu kṛṣikārye viśeṣataḥ |
kāryāntareṣv api grāmyeṣv evaṁ bhūpālakarmani ||215||
brāhmaṇāḥ kṣatriyā vaiśyāḥ śūdrāś ca dṛḍhabuddhayaḥ |
sarve sarvatra yoktavyāḥ tadā kṣemāya kalpate ||216||
tadā janapadasyāpi rāṣṭrasya ca viśeṣataḥ |
sukhādir aikamatyāc ca jāyate paramaṁ sukham ||217||
ekakaṇṭhyaṁ caikamatyaṁ pataṁgeṣu hi nityaśaḥ |
dṛśyatām iti nirdiṣṭaṁ bhārgaveṇeha mauninā ||218||
tasmāt grāme janapade pure vā durgakalpane |
nivāsino janāḥ sarve gatāsūyādidurguṇāḥ ||219||

setusaṁrakṣaṇe mārgarakṣaṇe kṛṣirakṣaṇe ।
puragrāmādirakṣāsu cākrīḍodyānarakṣaṇe ॥220॥
prajānāṁ rakṣaṇe cāpi gavām api ca rakṣaṇe ।
dharmarakṣāsv api tathā nyāyakāryasya rakṣaṇe ॥221॥
yac cānyat rakṣaṇaṁ proktaṁ nānārūpaṁ maharṣibhiḥ ।
sarvāsu tāsu rakṣāsu sarvaiḥ grāmādivāsibhiḥ ॥222॥
pratyekam ādaraḥ śreyān śasyate śāstracoditaḥ ।
yasmin grāme 'tha vā rāṣṭre rājñaḥ sadma na kalpitam ॥223॥
tasmin deśe viśeṣeṇa prajānāṁ pālanāya hi ।
amātyo vā rājavaṁśyaḥ guṇī pratinidhis tu vā ॥224॥
prakurvīta nivāsaṁ tu dhanasenāsahāyavān ।
prajābhiḥ prārthitaṁ datvā rakṣet tāś cāpado bhṛśam ॥225॥
nānāvidhair upāyais tu jīvanaṁ parikalpayet ।
brāhmaṇāḥ kṣatriyā vaiśyāḥ śūdrāḥ saṁkīrṇajātijāḥ ॥226॥
ye cānye manujā loke grāmādisthalavāsinaḥ ।
svadharmaniratās te tu sukhabhājo yaśasvinaḥ ॥227॥
saṁprāptaparamānandāḥ sakuṭuṁbāḥ sagodhanāḥ ।
virājante hi deśeṣu tatra tatra viśeṣataḥ ॥228॥
nṛpājñāvartinaḥ sarve śrutismṛtikṛtādarāḥ ।
paropakāraniratās tathā vītabhayā narāḥ ॥229॥
ādhivyādhivihīnās te satkriyāḥ sādhusevinaḥ ।
virājante hi deśeṣu tatra tatra viśeṣataḥ ॥230॥
sarveṣām eva teṣāṁ tu kṛṣivṛttiḥ sanātanī ।
śubhadā bhūtidā ślāghyā yathānyāyapravartanāt ॥231॥
brāhmaṇāḥ kṣatriyā vaiśyāḥ svakarmaniratā api ।
kṛṣikarmasu yogyās te sādhuślāghyaṁ hi tad viduḥ ॥232॥
yajñānām api cādhāraṁ prāṇināṁ jīvadāyakam ।
kṛṣikarma praśaṁsanti munayo divyacakṣuṣaḥ ॥233॥
nṛpāt prāptaṁ svataḥkrītaṁ sasyakṣetraṁ tu mānavāḥ ।

saṁprāpya yatnavantaś ca kṛṣīkāryakṛtādarāḥ ||234||
devānāṁ ca munīnāṁ ca te matāḥ prītidāyinaḥ |
dhanānām api sarveṣāṁ kṛṣir eva paraṁ dhanam ||235||
parairagrahyam ādiṣṭaṁ sarvaślāghyaṁ mahāphalam |
devānāṁ prītijanakaśuddhadravyapradāyi tat ||236||
pāratаṁtryaharaṁ caiva nityaṁ laksmīvilāsakṛtam |
tathātithīnāṁ devānāṁ svakuṭuṁbasya jīvadam ||237||
nānāvidhānandanakaraṁ kṛṣikarma praśasyate |
atas tad anyā vṛttis tu pāratantryeṇa guṁbhitā ||238||
puṇyānāṁ hānidā cāpi ninditā tattvadarṣibhiḥ |
ataḥ sadguṇasaṁpannāḥ puruṣāḥ svaprayatnataḥ ||239||
vedādhyayanasaṁpannā api rakṣāpravṛttayaḥ |
vānijyavṛttayaś cānye śūdrāḥ saṁkīrṇajā api ||240||
satyavratā dayāḍhyāś ca kṛṣyādānaratā bhṛśam |
saṁmānyavṛttayaḥ kāmaṁ sādhubhiś ca surair api ||241||
kṛṣipravṛttiṁ sarveṣāṁ devānāṁ prītidāyinīm |
yatnato rakṣayeyus tāṁ jīvānāṁ jīvanapradām ||242||
vane janapade deśe kṣetre grāmye bhṛgos taṭe |
kṛṣipravṛttiṁ saṁkalpyāṁ manyante hi munīśvarāḥ ||243||
purātanais tu munibhir aṭajāṅgaṇabhūmiṣu |
kṛṣikarmakṛtaṁ loke sarvaprāṇihitārthibhiḥ ||244||
tadādi kṛṣikāryaṁ tu khaṇḍeśaḥ sūkṣmabuddhibhiḥ |
nītaṁ deśeṣu sarvatra phaladaṁ cābhavat kramāt ||245||
ataḥ kṛṣyādānam etat sarvaiḥ puruṣapuṁgavaiḥ |
sūkṣmadhībhir ihāsevyaṁ śāśvatānandahetave ||246||

(6) kṛṣikāryādravyasaṁgrahaṇakathanam

kāśyapa upadiśati

medhāvī kṛṣikāryajñaḥ bhūlakṣaṇavidāṁ varaḥ |
śubhe muhūrte sumanasāstvabdāraṁbhe kṛṣikriyām ||247||

seveta phalasidhyarthaṁ trimūrtidhyānapūrvakam |
bhūdevīṁ dhānyadevīṁ ca smṛtvā manasi nirmale ||248||
pūjayitvātha vā bhaktyā dravyasaṁgrahaṇodyataḥ |
goṣṭhaṁ gosthānakaṁ vāpi gomayālepanakriyam ||249||
gandhapuṣpādyalaṁkārabhāsuraṁ ca viśeṣataḥ |
tindukais tiniśair vāpi sarjakaiḥ sāradārukaiḥ ||250||
gobandhatūlakaṁ kṛtvā śṛṁkhalādyair athāpi vā |
alaṁkṛtya tu taṁ deśaṁ gandhakuṁkumapuṣpakaiḥ ||251||
pradakṣiṇaṁ vā kṛtvā tatsthale gāś cārayet budhaḥ |
śubhalakṣaṇasaṁyuktāḥ dhenūḥ saṁcārayed iha ||252||
savatsāś cātha vṛṣabhān śvetāṅgān śubhalakṣaṇān |
suvarṇakupyaśṛṁgāgrān gandhādyaiś cāpy alaṁkṛtān ||253||
ghāsaiḥ śaṣpaiḥ palālaiś ca jalapātraiś ca maṁgalaiḥ |
pūrayed atha taṁ deśaṁ goṣṭhe śubhamuhūrtake ||254||
krītvā vā vṛṣabhān śvetān śubhalakṣaṇamānitān |
dhenūr vā mahiṣān vāpi bandhayet tatra kovidaḥ ||255||
lāṁgalān rajjujaṭilān halān api viśeṣataḥ |
sāṁgān nirīṣavantaś ca sāradārukṛtān tataḥ ||256||
śubhasthale 'tha nikṣipya gandhāmradalapuṣpakaiḥ |
candanaiḥ kuṁkumair vāpi maṇḍitāṁś ca viśeṣataḥ ||257||
saha maṁgalaghoṣeṇa dhānyadevīṁ ca pūjayet |
bhūdevīm atha lakṣmīm ca vāṇīṁ gaurīṁ surādhipam ||258||
varuṇaṁ ca dhaneśaṁ ca pūjayitvā viśeṣataḥ |
svakīyaṁ kṣetram āsādya vṛṣabhaiḥ saha kovidaḥ ||259||
lāṁgalair bhadrakair yuktaṁ sumuhūrte sulagnake |
karṣaṇaṁ kārayed ādau aiśānye koṇake śubhe ||260||
āgneye vāpi vā bhāge varuṇe vā kvacit sthale |
salilāpūraṇaṁ kṛtvā sukarṣāya tadā kramāt ||261||
prathamaṁ karṣaṇaṁ bhūmeḥ kṣetre svīye viśeṣataḥ |

kārayet vṛṣavāhyena halena kṛṣikāryavit ||262||
evaṁ krameṇa tat kṣetraṁ caturdhā paṁcadhāpi vā |
halena karṣayitvā ṣaṭdivasaṁ tv adhikaṁ tu vā ||263||
loṣṭhāṁś ca laguḍān anyān bahirnikṣipya gomayam |
ājakaṁ vālagaṁ tatra sthāpayet sāravṛddhaye ||264||
khanitrān śaṁkulān kṣudratalukān khaḍgakān tathā |
churikā yāś ca kṛṣyarhāḥ dravyaśāstreṣu niścitam ||265||
yuktyā tad ānayet sarvaṁ lavitraṁ lavasārakam |
yasmin deśe yadā vṛṣṭiḥ prathamā dṛśyate kṣitau ||266||
tadā bhūkarṣaṇaṁ śreṣṭham āhur atra munīśvarāḥ |
kulyāmukhena salilam āpūrya nijabhūmike ||267||
sasyārhakāle bhūmes tu karṣaṇaṁ ca munīritam
tasmāt sarvatra deśeṣu kṛṣikālānurūpataḥ ||268||
mitrāṇām upadeśāc ca yuktyā ca kṛṣikāryavit |
ādau bhūkarṣaṇaṁ kāryaṁ sunimitte sulagnake ||269||

(7) lāṅgalapūjākramakathanam

kāśyapa upadiśati

karṣaṇāt pūrvam eveha lakṣaṇajñaiḥ kṛṣīvalaiḥ |
lāṁgalādyarcanaṁ kāryaṁ vṛṣādīnāṁ viśeṣataḥ |270||
kṣetrasya vātha bhūmyāś ca pūjanaṁ phaladāyakam |
śuddhatoyena gandhādyaiḥ dīpadānair athāpi vā ||271||
bhūmidevi namas te 'stu mahi sarvaṁsahe 'dhunā |
kṛṣyārambhaṁ kariṣyāmi prasannā bhava suvrate ||272|
karṣaṇaṁ tāḍanaṁ yac ca tvayi yad yat kṛtaṁ mayā |
devi kṣamasva tat sarvaṁ kuru mahyaṁ mahāphalam ||273||
tvam eva mātā sarveṣāṁ prāṇinām iha kīrtyate |
ataḥ prasannā bhūdevi phalaṁ dehy amitaṁ kṣitau ||274||
iti stutvā prārthayitvā bhūdevīṁ viṣṇuvallabhām |
pradakṣiṇīkṛtya namaskārapūrvaṁ kṛṣīvalaḥ ||275||

dikdevān rakṣakāṁś cāpi sūryam āyuṣkaraṁ tathā |
stutvā bhūkarṣaṇavidhau kuryād ārambham uttamam ||276||

(8) tatra vṛṣabharājapūjākramanakathanam

kāśyapa upadiśati

saurabheya mahāsāra vṛṣarājāmitadyute |
bhūkarṣaṇavidhau tvaṁ hi sahyaṁ kuru mamānagha ||277||
sugandhamālyapuṣpādyair adya tvāṁ pūjayāmy aham |
phalado bhava me tvaṁ tu dharmakṛt svasti te bhṛśam ||278||
tṛṇaiḥ śaṣpaiś ca salilaiḥ tvāṁ poṣayāmy aham ādarāt |
śaṁkarānugrahāt tvaṁ tu phalado bhava me sadā ||279||
utpādya tanayān śaśvat tvattulyabalavīryakān |
alaṁkuru tvaṁ goṣṭhaṁ me tvāṁ poṣyāmy aham ādarāt ||280||
vṛṣarāja tvam evātra dhanadhānyādivṛddhikṛt |
dharmarūpa tvam eveha tasmāt tvāṁ poṣayāmy aham ||281||
devayajñaṁ bhūtayajñaṁ yathā me saphalaṁ bhavet |
tathā dayāṁ kuru tvaṁ tu cāparādhaṁ kṣamasva me ||282||
iti samprārthya vṛṣabhaṁ pūjayitvā viśeṣataḥ |
śrāntā yathā te na klāntās tāvat tatkāryam īritam ||283||
tanmanaḥkleṣahetus tu viphalāya viniścitaḥ |
tasmān na yojayet śrāntān vṛṣabhān kṛṣikarmani |
tadā gopūjanaṁ śastam ity āhur munisattamāḥ ||284||

(9) saṁgraheṇa govṛṣabhalakṣaṇakathanakramaḥ

kāśyapa upadiśati

gāvaś ca vṛṣabhāś caiva śubhalakṣaṇabhāsurāḥ |
śubhadāḥ svāmināṁ proktāḥ deśakṣemapradāś ca te ||285||
kṛṣikarmaṇi te yogyāḥ vṛṣabhā vṛṣavardhakāḥ |
śvetāṁgāḥ kiṁcid uttuṁgāḥ vṛṣabhā maṁjunetrakāḥ |
gaṁbhīraninadā ye ca te tu brāhmaṇajātikāḥ ||286||
raktāṁgā nitarāṁ tuṁgāḥ gaṁbhīraninadāś ca ye |

tejobalasametāś ca te vṛṣāḥ kṣatrajātijāḥ ||287||
śvetaraktāṁkanair yuktāḥ tanmaṇḍalavibhūsitāḥ |
nātituṁgā nātinīcāḥ te vṛṣā vaiśyajātijāḥ ||288||
kṛṣṇāṁgā nātituṁgāś ca kopavegena tāḍitāḥ |
su++++vantaś ca te vṛṣāḥ śūdrajātijāḥ ||289||
ete sarve praśastāś ca kṛṣikarmaṇi yogyakāḥ |
nātidīrghaṁ na ca kṛiśaṁ vṛṣaśṛṁgadvayaṁ śubham ||290||
mithovaiṣamyahīnaṁ ca vṛṣaśṛṁgaṁ praśasyate |
khurāś ca vṛṣabhānāṁ tu na kṛśā nātidīrghakāḥ ||291||
vaiṣamyahīnāś ca tathā dṛḍhāḥ ślāghyā viniścitāḥ |
śubhā gatir dīrghavālaḥ vṛṣabhānāṁ praśasyate ||292||
nātisthūlā na ca kṛśāḥ bhāravāhitvam eva ca |
prasannatvaṁ kāryakāle vṛṣāṇām uttamā gunāḥ ||293||
khure viṣāṇe śvetaś ca lalāṭe pūcchake 'pi ca |
varṇāntaravibhūtiś ca vṛṣaḥ śubhavivṛddhidaḥ ||294||
ramyātmānaś ca hṛṣṭāś ca śubhavarṇasamujvalāḥ |
śubhāvartādisahitāḥ dhanadhānyavivṛdhhidāḥ ||295||
sūkṣmaromāvalīvyāptāḥ gaṁbhīraninadāś ca ye |
manojñanayanopetāḥ vṛṣabhāḥ śubhadāḥ matāḥ ||296||
raktavarṇā mahāśṛṁgāḥ tejobalavivarjitāḥ |
skhaladgatiyutāś cāpi krodhanidrāvaśaṁgatāḥ ||297||
kṛśāṁgā dhūmravarṇāś ca tathā bahvāśinaś ca ye |
ye vivarṇāś ca vṛṣabhāḥ nitarāṁ nīcakāś ca ye ||298||
karkaśāṁgāḥ krūraravāḥ varjyās te vṛṣabhādhamāḥ |
saṁpūrṇaśvetakāyāś ca raktakāyās tu vā vṛṣāḥ ||299||
śubhadā dhanadhānyādivṛddhidāḥ parikīrtitāḥ |
doṣahīnā guṇopetāḥ vṛṣabhāḥ śuddhajātijāḥ ||300||
saṁgrāhyāḥ kṣemasiddhyarthaṁ kṛṣikārair viśeṣataḥ |
sulakṣaṇā dhenavaś ca mahiṣāś ca tathā matāḥ ||301||

mahiṣyāś ca bahukṣīrāḥ rakṣanīyāḥ kr̥ṣīvalaiḥ ।
meṣāḥ chāgāś ca bahudhā kr̥ṣikāryaphalapradāḥ ॥302॥
paraṁparopadeśena parīkṣāsv api paṇḍitaiḥ ।
niścitān doṣahīnāṁś ca vr̥ṣabhādīn śubhārthinaḥ ॥303।
rakṣeyuś ca prayatnena kr̥ṣisāphalyahetave ।
kāleṣu poṣanīyāś ca hitāhārapradānataḥ ॥304॥
rogebhyo rakṣanīyāś ca lālanādibhir anvaham ।
goṣṭhasthalaṁ nirdiśec ca teṣāṁ puṣṭipradaṁ budhaḥ ॥305॥
kr̥ṣikarmaṇi cādyaṁ tu sahāyā vr̥ṣabhottamāḥ ।
ataḥ satkulasaṁjātā dhenavaś ca viśeṣataḥ ॥306॥
kr̥ṣīvalais tatra tatra grāmeṣu nagareṣv api ।
śubhalakṣaṇasaṁpannāḥ dhenavas tu paraḥśatam ॥307॥
pālanīyāḥ kr̥ṣikaraiḥ grāmeṣu nagareṣv api ।
palālatr̥ṇanīrādyaiḥ lālanair bahukrakṣaṇaiḥ ॥308॥
dharmakāmārthasiddhis tu dhenūnāṁ pālanāt bhavet ।
rogapraśāntiś ca tathā bālavr̥ddhādirakṣaṇam ॥309॥
gosahasraṁ yatra deśe pālyate prītipūrvakam ।
tatra deśeṣu parjanyo varṣaty eva na saṁśayaḥ ॥310॥
gavāṁ rakṣaṇato dānāt devānāṁ prītir uttamā ।
ataḥ prajānāṁ kṣemādir ity evaṁ bhārgavo ’bravīt ॥311॥
ataḥ sarvaprayatnena gogaṇaṁ tu kr̥ṣīvalāḥ ।
rakṣeyuś ca vanānteṣu grāmamadhye ca sarvataḥ ॥312॥
svayaṁ bhr̥tyagaṇaiś cāpi gopālaiś ca viśeṣataḥ ।
pratyabdaṁ tu śubhe kāle gopūjā ca viśiṣyate ॥313॥
dinānte pratyahaṁ tāsāṁ gandhapuṣpādipūjanam ।
yaśasyaṁ subhadaṁ proktaṁ sarvaśāstreṣu kīrtitam ॥314॥
satsantānapradaṁ puṁsāṁ trivargaphaladaṁ tathā ।
chāyāsu tāś ca te chāgāvr̥ṣabhādyāḥ kriyākarāḥ ॥315॥
madhyāhvakāle grīṣme tu sthāpyā rakṣyāś ca dhīmatā ।

śītakāleṣu goṣṭhe tu hitāhārārpaṇādibhiḥ ||316||
saṁrakṣyāḥ kṣemasiddhyarthaṁ kuṭuṁbasyāpi bhūpateḥ |
gorakṣaṇaṁ tu bhūpālaiḥ kartavyaṁ dharmyam īritam ||317||
corāc ca duṣṭasatvāc ca tadrakṣaṇam athārpaṇam |
mahāphalaṁ samuddiṣṭaṁ kṛṣikārye viśeṣataḥ ||318||
iti śāstropadeśakramaḥ
atha prathamo bhāgaḥ
dhānyādikṛṣikramakathanam
(10) sasyārhakālaparīkṣākramaḥ
kṣetravibhāgakramaś ca
kāśyapa upadiśati
kṛṣikāryakramavido loke sarvatra dhījuṣaḥ |
vṛṣabhān mahiṣān dhenūḥ bhṛtyān api ca sevakān ||319||
poṣayitvā niyogajñān kṛṣiṁ kṣetreṣu kalpitān |
vardhayeyuḥ kramamukhaṁ vividhaṁ kṣetrayogyataḥ ||320||
ādau kālaparīkṣātra mukhyakalpā vidhīyate |
kāśmīradeśyake vaṁgadeśye nepāladeśyake ||321||
pāṁcālakosalakuruvirāṭāvantibhūmike |
mālave śakadeśye vā sindhusauvīrabhūmiṣu |
śūrasenāvanticedikoṁkaṇāndhrādibhūmiṣu ||322||
yadā yadā vṛṣṭipātas tadā śastā kṛṣikriyā |
yatra grīṣmādikāleṣu nadīkulyādimārgataḥ ||323||
kṣetreṣu salilotkṣepaḥ tadā śastā kṛṣikriyā |
yatrāgādhajalādhārapraṇālīkulyakāmukhāt ||324||
yatra yatra jalotkṣepaḥ tadā tatra kṛṣiḥ śubhā |
meghodayajalasrāvam agādhasalilāśayam ||325||
kulyājalasamṛddhiṁ ca vīkṣya dhīmān kṛṣīvalaḥ |
kṛṣiṁ prakalpayet yuktyā sthalasaṁpatkramād bhuvi ||326||
dvidhā vibhaktāni loke kṣetrāni svāyarūpataḥ |

tatra tatra naditīre grāmeṣv api vanāntare ||327||
parvatasya taṭe caivaṁ nimnonnatatale 'pi ca |
śālihūmis tu tatrādyā munibhiḥ kathitottamā ||328||
dvitīyāḍhakabhūmis tu madhyamāpi phalapradā |
śālibhūmis tu salilapracurā tu svabhāvataḥ ||329||
saṁpūrṇakardamā khyātā nitarāṁ mr̥dumr̥ttikā |
kulyopakulyā śākhāliguṁbhitā tajjalasravā ||330||
samantataḥ kṣudrarūparakṣāmr̥ttikabhittikā |
khalasvarūpato vāpi grāmavāstusvarūpataḥ ||331||
kṣetrapaṁktisvarūpād vā śālibhūmisthitir matā |
prācyāṁ kvacin nimnabhāgā prāticyām atha ca kvacit ||332||
yāmye kvacin nimnatalā kauberyāṁ nimnabhūmikā |
kvacit prāyaḥ samatalā kulyā nīragrahonmukhī ||333||
kṣetrāt kvacit kṣetratalapraveśārhajalasravā |
evaṁ svarūpaṁ vividhaṁ śālibhūmyās tu niścitam |
prathamakarṣaṇādīha vrīhyutpacyavadhi kṣitau ||334||
śālibhūḥ pūrṇasalilā khyātā pūrṇaphalapradā |
grāmādivastubhūmyāṁ ca vanānte ca taṭe kvacit ||335||
unnatā bhūr dvitīyā sā kathitāḍhakabhūmikā |
svalpodabindusaṁsevyā caṇakāḍhakavardhinī ||336||
naitasyāḥ jalakulyādiprakalpanam udīritam |
prāyeṇa svalpasalilasekāt karṣaṇayogyakā |
tato bījāvāpakāle kiṁcit salilapānatā ||337||
evaṁ samucite kāle jīvanaṁ bījajīvadam |
etādr̥śaguṇopetā kathitā cāḍhakasthalī ||338||
śālibhūmyāś cāḍhakādibhūmyāś cārhaṁ kr̥ṣīvalaiḥ |
bījasaṁgrahaṇaṁ nānārūpam atra munīritam ||339||

(11) nānāvidhibījasaṁgrahaṇakramakathanam

kāśyapa upadiśati

pūrvaṁ saṁprārthito dhātryā dhātā lokapitāmahaḥ ǀ
bījāni vividhānīhāsṛjat punar acintayat ǀǀ340ǀǀ
tadāśayaṁ viditvā tu bhūmidevī mahodayā ǀ
uvāca harṣeṇa vidhiṁ prāṇināṁ sukhasiddhaye ǀǀ341ǀǀ
vidhe nanu tvayā sṛṣṭānīmāni vividhāni tu ǀ
bījāni mayi nikṣiptāny ādarād acireṇa hi ǀǀ342ǀǀ
prāptāṁkurāni niyataṁ bhaviṣyanti na saṁśayaḥ ǀ
tavānugrahabhūmnā ca vāsavasya viśeṣataḥ ǀǀ343ǀǀ
pratyabdaṁ meghavṛṣṭiś ca niyatena bhaviṣyati ǀ
yathākālaṁ prāptavarṣāḥ prajāḥ sarvatra sarvataḥ ǀǀ344ǀǀ
kṛtvā bījāvāpanaṁ tu sukṣetreṣu viśeṣataḥ ǀ
saṁpūrṇaphalabhājaś ca puṣṭā hṛṣṭāś ca nityataḥ ǀǀ345ǀǀ
bhaviṣyanti kṛṣijñāś ca nānādeśeṣu bhūtale ǀ
devānāṁ prītiratulā yajñādīnāṁ vivardhanam ǀǀ346ǀǀ
prajānām atithīnāṁ ca prīṇanaṁ bhojanādibhiḥ ǀ
dhenvādiprīṇanaṁ kālyaṁ saphalaṁ bhūtayajñakam ǀǀ347ǀǀ
iti vijñāpito devyā medinyā kamalāsanaḥ ǀ
tathāstv ity abravīd vākyaṁ sarvaprāṇihitapradam ǀǀ348ǀǀ
tadādi loke sarvatra bījāni vividhāni tu ǀ
rakṣyante kṛṣikarmajñaiḥ mānavaiḥ sūkṣmabuddhibhiḥ ǀǀ349ǀǀ
śālyādikas tu prathamaḥ nirdiṣṭo hi munīsvaraiḥ ǀ
āḍhakādir dvitīyaś ca tṛtīyaḥ śākasantataiḥ ǀǀ350ǀǀ
latākusumavargas tu caturthaḥ kila kīrtitaḥ ǀ
etat caturvidhaṁ loke kṛṣikarma viniścitam ǀǀ351ǀǀ
puṣṭipradaṁ ca phaladaṁ mānavānāṁ viśeṣataḥ ǀ
sarvartuṣv api santoṣajanakaṁ devatuṣṭidam ǀǀ352ǀǀ
govṛṣoṣṭrāśvakuṁbhyādijīvināṁ jīvadāyakam ǀ
nāradādyair yogivaryair upadiṣṭaṁ hi bhūtale ǀǀ353ǀǀ
mahīpālair uttamaiś ca svasvadeśasthalīṣv api ǀ

kr̥ṣikarma kr̥taṁ bhaktyā bhr̥tyavargaiś ca rakṣitam ||354||

tadādi brāhmaṇavaraiḥ kṣatriyaiḥ vaiśyajātijaiḥ |

śūdrair anyaiś ca bahudhā sukṣetre kālayogyakam ||355||

ārabdhaṁ kr̥ṣikāryaṁ tu bījamukhyaṁ prakīrtitam |

ataḥ sudhībhir bhūloke kr̥ṣikarmavidāṁ varaiḥ ||356||

bījānāṁ saṁgraho yuktyā kalpanīyo viniścitaḥ |

nāmāni kramaśo vakṣye bījānām iha siddhaye ||357||

śālyādi kalamādiś ca ṣaṣṭikādiś ca tattridhā |

rasavarṇakajātyādyaiḥ vibhaktaṁ kr̥ṣikovidaiḥ ||358||

kiṁcit ghanās tu kalamāḥ rasādhikyāḥ samujvalāḥ |

ṣaṣṭikā rasahīnāś ca kr̥tāḥ kamalayoninā ||359||

śālivrīhis tu ṣaṭviṁśat bhedakaḥ parikīrtitaḥ |

tatra tatra hi deśeṣu bhūmibhedena kovidaiḥ ||360||

bījāvāpaḥ prakartavyo yathākālam atandritaiḥ |

tasmāt sarvaprayatnena tadrakṣaṇam udīritam ||361||

śvetaśāliḥ raktaśāliḥ sthūlaśālis tathaiva ca |

dīrghaśāliḥ svādurasaḥ kalamaḥ śvetavarṇakaḥ ||362||

raktavarṇaś ca kalamaḥ kalamaḥ sthūladehakaḥ |

dīrghākr̥tiś ca kalamaḥ vrīhir hemākhyaśaṁbakaḥ ||363||

kapiśaḥ śaṁbako 'nyaś ca raktaśaṁbaś ca varṇataḥ |

kr̥ṣṇaśaṁbaś cānyarūpaḥ nitarāṁ rasavanmataḥ ||364||

śukavrīhiḥ sthūlakāyo vrīhiś cānyaḥ prakīrtitaḥ |

ghanavrīhiś ca palaśavrīhiś ca rasameduraḥ ||365||

svāduvrīhiś ca nitarāṁ phalavr̥ihis tathāparaḥ |

drākṣāvrīhiś cāparas tu nīvāraḥ śvetakr̥ṣṇakaḥ ||366||

yavau ca śvetakr̥ṣṇau ca sthūlākāras tathaiva ca |

saṁmaravrīhir aparaḥ nitarāṁ rāśikārakaḥ ||367||

kalavrīhiś ca paramaḥ nitarāṁ puṣṭidāyakaḥ |

sitavrīhiḥ pītavarṇaḥ vrīhiś cājīrṇahārakaḥ ||368||

iti ṣaṭviṁśatibhidākalpitā dhānyarāśayaḥ ǀ
sarveṣām api caiteṣāṁ bījānāṁ rakṣaṇaṁ kramāt ǁ369ǁ
arkātape śoṣitānāṁ supakvānāṁ khalasthale ǀ
svagrahe vā rakṣaṇaṁ tu pradhānaṁ śubhavardhakam ǁ370ǁ
bhūpālair athavā rakṣyaṁ kāle deyaṁ prakīrtitam ǀ
bījarāśiḥ paraśreyaḥ kṛṣes tu prathamaṁ dhanam ǁ371ǁ
narāṇām api devānāṁ prītidaṁ dānayogyakam ǀ
tasmāt kṛṣīvalair yuktyā bījasaṁrakṣaṇaṁ varam ǁ372ǁ
etāni tāni bījāni saṁpūrṇaphaladāni ca ǀ
māseṣu triṣu vā tāni caturmāse 'tha vā sthale ǁ373ǁ
kānicit paṁcamāseṣu ṣaṇmāseṣv iha kāni ca ǀ
saptamāseṣv aṣṭamāseṣv adhikakramataḥ kvacit ǁ374ǁ
phaladāni supakvāni nirṇītāni munīśvaraiḥ ǀ
tasmāt svadeśe kṣetreṣu jalasrāvānukūlataḥ ǁ375ǁ
śālyādikṛṣikāryeṣu cārambhaḥ sukhasiddhaye ǀ
pūrvapuruṣakāryādikramatas tu kvacit sthale ǁ376ǁ
meghodayajalasrāvakramataś cānyato bhuvi ǀ
deśakālānuguṇyena kṛṣikarma praśasyate ǁ377ǁ
narendrāṇāṁ prajānāṁ ca phaladaṁ parikīrtitam ǀ
dhenvādinām api tathā pataṁgānām api kramāt ǁ378ǁ
jīvasukhyapradam idaṁ kṛṣikarmasuniścitam ǀ
evam āḍhakabījāni māṣabījāni ca kramāt ǁ379ǁ
mudgānāṁ caṇakānāṁ ca godhūmānāṁ viśeṣataḥ ǀ
tīlānāṁ kīrakānāṁ ca kodravānāṁ kvacit sthale ǁ380ǁ
yavarānāṁ kulasthānāṁ śyāmānāṁ kṛṣasāriṇām ǀ
ikṣūnām api puṇḍrāṇāṁ jātibhedarasakramaiḥ ǁ381ǁ
vibhaktānāṁ bahuvidhaṁ dhānyānām uṣṇabījakam ǀ
jīrakānāṁ sarṣapānāṁ marīcīnāṁ tathaiva ca ǁ382ǁ
vidheyānāṁ ca tūlānām eraṇḍānāṁ kvacit sthale ǀ

eteṣām api bījānāṁ rakṣaṇaṁ sukhadāyakam ||383||
paṭolikānāṁ bījāni vārtākānāṁ tathaiva ca |
jaṭikā rāśijaṭikā vallikā vanavallikā ||384||
ṣavakā cāpi vividhā rasapūrṇā ca kīrtitā |
śākāś ca vividhā varṇarasajātiprabhedakāḥ ||385||
kūṣmāṇḍānāṁ kalāṭānāṁ nānārūpaṁ tu bījakam |
kustuṁbānāṁ kaṭānāṁ ca dhānyakānām api kvacit ||386||
haridrānāṁ ca śuṇṭhīnāṁ bījasaṁrakṣaṇaṁ śubham |
sūraṇānāṁ ca kandānāṁ śākuṭānāṁ tathaiva ca ||387||
rasakānām api tathā kadalīnāṁ viśeṣataḥ |
bījasaṁrakṣaṇaṁ kāryaṁ kr̥ṣikarmaviśāradaiḥ ||388||
elā drākṣā ca naladaṁ nāgavallī rasojvalī |
pūgānāṁ kramukānāṁ ca bījasaṁrakṣaṇaṁ śubham ||389||
mallikāpītapuṣpānāṁ kundānām api bījakam |
rakṣeyur atimuktānāṁ cāṁpeyānām api kvacit ||390||
śigrūṇāṁ bhūrjakānāṁ ca karaṁjānām api kvacit |
śamīnāṁ naktamālānāṁ devadārvākhyabhūruhām ||391||
vibhītakapriyaṁgvoś ca tathāmalakabhūruhām |
piṇḍītakānāṁ niṁbānāṁ likucāṇām api kvacit ||392||
śirīṣāṇām aśokānāṁ dāḍimānāṁ tathaiva ca |
śrīparṇānāṁ tamālānāṁ sindhuvārākhyabhūruhām ||393||
kapitthānāṁ ca jaṁbūnāṁ saptaparṇākhyabhūruhām |
bījasaṁrakṣaṇaṁ kāryaṁ sukhāya ca kr̥ṣīvalaiḥ ||394||
uduṁbaraś caladalāḥ tiniśāḥ pāribhadrakāḥ |
palāśāś ca madhūkāś ca vaṭāḥ plakṣā rasālakāḥ ||395||
tindukās tilakā nīpāś cārjunāḥ sarjakāśca ye |
badarāḥ pīluvr̥kṣāś ca jaṁbīrāḥ punnagāś ca ye ||396||
vaṁśāḥ kāśāś ca vividhāḥ mātulaṁgākhyābhūruhaḥ |
evaṁ tu vividhā vr̥kṣāḥ patraiḥ puṇyaiḥ phalair api ||397||

phalapradā mānavānāṁ pataṁgānāṁ viśeṣataḥ ।
vaṭādayo mahāvṛkṣāḥ ye cānye parikīrtitāḥ ॥398॥
sarveṣām eva teṣāṁ tu bījasaṁrakṣaṇaṁ hitam ।
supakvānām ātape tu śoṣitānāṁ viśeṣataḥ ॥399॥
bījānāṁ santatī rakṣyā bhāṇḍeṣu vividheṣu ca ।
palālarāśiṣu tathā kuṇḍeṣv api kṛṣīvalaiḥ ॥400॥
śaśamūṣakamārjāraiḥ na dūṣyā bījasantatiḥ ।
jalarāśikaṇair vāpi vātavṛṣṭyādibhis tathā ।
yathā na dūṣitā doṣairanyaiś ca vividhair api ॥401॥
bījānāṁ santatiḥ pālyā tathā gehe khalasthale ।
bījānām uttamānāṁ tu pālanaṁ dharmabuddhitaḥ ॥402॥
śreyase kṛṣikānāṁ tu kathitaṁ munipuṁgavaiḥ ।
śālyādibījāni tathā śākabījāni vā punaḥ ॥403॥
vṛkṣāṇām api bījāni kundānām api bījakam ।
krītvā saṁgrahya vā loke kṛṣikarmavido narāḥ ॥404॥
brāhmaṇāḥ kṣatriyā vaiśyāḥ śūdrāḥ saṁkīrṇajā api ।
svakṣetreṣu svadeśastheṣv ādarād yuktitas tathā ॥405॥
nānārūpeṣu vṛṣabhaiḥ karṣiteṣu viśeṣataḥ ।
kāle samucite siktāny atha vā vardhitāni ca ॥406॥
āvāpavidhinā yuktyā vinyaseyur yathākramam ॥407॥

(12) bījāvāpārthaṁ bhūkarṣaṇakramaḥ

kāśyapa upadiśati

bījāvāpas tu vividho nirdiṣṭaḥ śāstrakovidaiḥ ।
kvacit śāleyabhūmyāṁ tu karṣaṇaṁ vṛṣābhaiḥ śubhaiḥ ॥408॥
ato nirdiṣṭam eteṣāṁ bījānāṁ vrīhisantateḥ ।
kandādīnām api tathā vardhanaṁ prathamaṁ sthale ॥409॥
kāryam ity evam ādiṣṭaṁ kṛṣikarmavicakṣaṇaiḥ ।
jalakṣetre ca tadbhinne jalasikte śubhasthale ॥410॥
bhūkarṣaṇaṁ ca vṛṣabhaiḥ kārayitvā yathākramam ।

bījāvāpanayogyāni kṣetrāṇi vividhāni tu ||411||
kārayitvā samīkṛtya cāṁkuranyāsanaṁ tu vā |
kalpayeyur yathāśakti yuktyā ca sthalayogyakam ||412||
ākrīḍe cāpi cārāme nṛpodyāne tathaiva ca |
grāmasya nagarasyāpi bahir antas tu vā sthale ||413||
vṛṣabhair bhūkarṣaṇaṁ ca khanitrādyaiḥ samakriyā |
kheṭādyaiś ca samīkāraḥ kṣetrāṇāṁ parikīrtitaḥ |
sītākramaś ca kṣetreṣu bījāvāpāya kovidaiḥ ||414||
kvacit kalpaś ca nirdiṣṭaḥ kvacit khananam eva ca |
kandādīnāṁ sthāpanārtham avaṭaṁ vā prakalpayet ||415||
kadalyādyā ye ca latātaravaḥ saṁprakīrtitāḥ |
teṣāṁ bījāṁkuraṁ yuktyā kandasthāpanam eva ca ||416||
latayā vardhanaṁ cāpi vividhaṁ kṛṣikovidaiḥ |
samīkṛte hi sukṣetre yathākāle sapeṣṭake ||417||
nyasanaṁ phaladaṁ coktam ṛṣibhis tattvadarśibhiḥ |
tasmāt bhūkarṣaṇaṁ śastaṁ śālikṣetre latāsthale ||418||
śākādivṛddhyarhabhūmau udyāne vṛkṣakasthale |
karṣaṇaṁ cāvaṭaṁ sītārekhākalpanam eva ca ||419||
tattatkṣetrānukūlyena kalpayet kṛṣikovidaḥ |
jalasekāt tena hīnam arkarocichaṭākramāt ||420||
śuṣkitaṁ vā sthalaṁ dhīmān bījāvāpāya nirdiśet |
latādisthāpane caivaṁ vṛkṣāṇāṁ sthāpane tathā ||421||
kadalīkandavinyāse hy aṁkurasthāpane 'pi ca |
bhūkarṣaṇādikaṁ sarvaṁ kṛṣikāryaṁ tu yuktitaḥ ||422||
paraṁparopadeśāc ca kārayet phalasiddhaye ||423||

(13) vividhadhānyaprāptyarthaṁ kṛsikramādikathanam

kāśyapa upadiśati

kośalādiṣu deśeṣu nadīsalilapūrite |
sasyakṣetrasamūhe tu śālyādiḥ kṛṣir uttamā ||424||

nirdiṣṭā munibhis tasmāt dhīmadbhis tu kṛṣīvalaiḥ ǀ
kalamavrīhinivahān prāptuṁ kṛtaviniścayaiḥ ǀǀ425ǀǀ
jalāśayānām athavā nadīnāṁ ca hradasthale ǀ
kulyāmukhāt yathākālaṁ sasyakṣetreṣu coditam ǀǀ426ǀǀ
ādāvāpūrya tadanu karṣaṇaṁ vṛṣabhair matam ǀ
tataḥ salilapūrṇāni sasyakṣetrāṇi mānavāḥ ǀǀ427ǀǀ
śāṁkarāt dikoṇabhāgād athavā tu kvacit sthale ǀ
prācyāṁ pratīcyām athavā karṣaṇaṁ śubhadāyakam ǀǀ428ǀǀ
pūrvam eva kvacit deśe nibiḍaṁ vardhitāni ca ǀ
śālyaṁkurāṇāṁ jālāni cotkhātakramam ādiśet ǀ
bandhīkṛtya tu tajjālaṁ karṣaṇān mṛdutāṁ gate ǀǀ429ǀǀ
jalakardamabhūyiṣṭhe śālikṣetre sulagnake ǀ
ajāśakṛtkaṇair evaṁ gavāṁ cāpi śakṛtkaṇaiḥ ǀǀ430ǀǀ
latāvratatibhir vāpi prāptasāre sayatnake ǀ
paṁktiśaḥ paṁktiśo bhṛtyaiḥ vinyaset samabhūmike ǀǀ431ǀǀ
tatra kṣetre svalpajalaṁ sthāpanīyam anantaram ǀ
jalādhikyasrāvamārgaḥ kalpanīyo viśeṣataḥ ǀǀ432ǀǀ
yathā sasyāṁkurāṇāṁ tu nāśas tu na bhavet tathā ǀ
tathā tadrakṣaṇaṁ kāryaṁ tatkarmaphaladaṁ viduḥ ǀǀ433ǀǀ
evam aṁkurajālānāṁ vinyāsaś ca kramān mataḥ ǀ
yatrāṁkurāṇāṁ nāśas tu vṛṣṭyādibahukāraṇaiḥ ǀǀ434ǀǀ
saṁpāditas tathā dhīmān tatra deśe prayatnataḥ ǀ
navāṁkurāṇāṁ nyasanaṁ phaladaṁ parikalpayet ǀǀ435ǀǀ
grāmyeṣu kṣetrajāleṣu vanye vātha sthalāntare ǀ
ekadvitricatuḥpaṁcakuṭuṁbibhir udasthale ǀǀ436ǀǀ
daśakair viṁśatiguṇaiḥ vṛṣabhair yugapaṁktibhiḥ ǀ
ekadā sasyabhūmīnāṁ karṣaṇaṁ sasyavṛddhikṛt ǀǀ437ǀǀ
ekamatyāt grāmajanaiḥ bahuyogyaprayogataḥ ǀ
ekadā bījavinyāsaḥ phaladaś ca prakīrtitaḥ ǀǀ438ǀǀ

evaṁ sakardamajale śālikṣetre viśeṣataḥ |
grāmye vā nagare vāpi vanye vanatalāntare ||439||
sakulyājalanisrāve śālyaṁkuragaṇe 'rpite |
jalādhikakṣaye tatra kalpite bahuvaktrake ||440||
kṣetranyastā śālibījāṁkurapālī tu bhūtale |
dināni saptadaśa vā tīrtvā sākr̥ṣṭabhūmike ||441||
saṁprāptamūlā vr̥ddhyarthaṁ brahmaṇā tu vidhīyate |
tato bahir navadalaprāptiś ca śubhadarśanaṁ ||442||
yā kāntiḥ śukapatreṣu dr̥śyate sumanoharā |
tayā kāntyāṁcitā bhāti bālevārūḍhayauvanā ||443||
evaṁ tu prathame māse gate vrīhisthale bhuvi |
kalamādidalānāṁ hi vr̥ddhiḥ saṁdr̥śyate kramāt ||444||
paritoṁ 'kuravr̥ddhiś ca tatsamā kāntibhūṣitā |
pratyahaṁ pītatoyā ca phaladānāya kalpate ||445||
sasyakṣetrasamūheṣu tadānīṁ tr̥ṇakoṣṭhakāḥ |
sakardamajaleṣv atra vr̥ddhibhājaś ca niścitāḥ ||446||
ta eva sasyaripavo niścitā munisattamaiḥ |
tasmāt kr̥ṣīvalair yuktyā tadā tu tr̥ṇakoṣṭhakāḥ ||447||
muṁjādyāś cānyadoṣā ye nirasyā yatnato bhuvaḥ |
muṁjākoṣṭhatr̥ṇādīnāṁ phalahānipradāyinām ||448||
sasyavr̥ddhikṣayakr̥tāṁ mūlataś chedanaṁ śubham |
śāleyabhūmiṣu tadā jalamāpūrya kovidaḥ ||449||
tr̥ṇakoṣṭhān nirasyātha paṁktiśaḥ paṁktiśaḥ kramāt |
bhr̥tyavargaiḥ pratyahaṁ vā vairichedaḥ praśasyate ||450||
sasyakṣetreṣu sarvatra caivaṁ rītyā tr̥ṇādiṣu |
dviṣatsu saṁnirūddheṣu yatnatas tu kr̥ṣīvalaiḥ ||451||
prāptaśobhā vr̥ddhiyutā bhāsate śālisantatiḥ |
tato dvitīyamāse tu śālivr̥ddhiś ca niścitā ||452||
tadānīm api ye kecit tr̥ṇādiripavo yadi |

bhavanti sasyakṣetreṣu tannirāsaś ca yuktitaḥ ||453||
mahāphalāya kathitaḥ sasyānāṁ bahurūpiṇām |
śvetaśālī raktaśāliḥ kalamādyāś ca ye purā ||454||
nirdiṣṭā vividhā dhānyarāśayo bahujātikāḥ |
sarveṣām api teṣāṁ tu kṛṣikāryaṁ hi tulyakam ||455||
bhūmīnāṁ karṣaṇaṁ cādau tato bījāṁkurārpaṇam |
tṛṇakāṣṭhotkṣepaṇaṁ ca punas tatkarma cādarāt ||456||
jalasaṁpūraṇaṁ caiva śālisaṁrakṣaṇaṁ tathā |
tulyarūpaṁ tulyakāryaṁ niścitaṁ kṛṣikovidaiḥ ||457||
tathāpi śālayaḥ śaṁbāḥ kalamādyāś ca bhūmiṣu |
jātibhedāt varṇabhedāt trimāsaphaladās tu te ||458||
kecic caturmāsaphalāḥ paṁcamāsaphalāḥ pare |
ṣaṇmāsaphaladāś cānye tadādhikyaphalapradāḥ ||459||
tasmāt taddhānyakaṇikāprādurbhāvakramaṁ bhuvi |
jñātvā kṛṣīvalais tāvat teṣāṁ salilapūraṇam ||460||
nirdiṣṭaṁ kṛṣikarmajñaiḥ deśe sarvatra sarvataḥ |
yadā tu dhānyakaṇikarāśiḥ sasyāgrapaṁktiṣu ||461||
kṣīragarbho ghanaḥ kiṁcid ānataḥ dṛśyate kramāt |
tadā tatkaṇikārāśirakṣaṇaṁ tu śukād api ||462||
tulyam eva vinirdiṣṭaṁ nānādeśeṣv apīritam |
evam ataḥ kṣīrapūrṇā dhānyānāṁ kaṇikāvaliḥ ||463||
kramād antaḥ sāravatī tatsāras taṇḍulodayaḥ |
tāvat paryantam eteṣāṁ salilasya prapūraṇam ||464||
śastaṁ ca phaladaṁ proktaṁ anyathā phalahīnatā |
tasmāt vrīhyantarasakṛtkṣīravṛddhikramādikam ||465||
jñātvā tatsamaye mūle sasyānāṁ kṣīravṛddhaye |
krameṇa salilasrāvaḥ kartavyas tu kṛṣīvalaiḥ ||466||
nadījalāśayajalapūraṇaṁ vā kvacit sthale |
hradakūpādisalilapūraṇaṁ vā prayatnataḥ ||467||

tadā phalapradaṁ proktaṁ prāṇināṁ sukhasiddhaye ǀ
vasundharāyā varataḥ saṁkalpāt brahmaṇas tathā ǀǀ468ǀǀ
uptāni sasyakṣetreṣu bījāni vividhāni tu ǀ
phaladāni bhavantīha vṛddhaṁ prāpya kramāt bhuvi ǀǀ469ǀǀ
svayam evaṁ vidhāny eva bhāsurāṇi mahītale ǀ
mūṣakāt śalabhakīrāt śukād api ca duṣṭataḥ ǀǀ470ǀǀ
prāṇivargāt rakṣaṇaṁ tu viśeṣaphaladaṁ viduḥ ǀ
parasparaṁ snehavantaḥ grāmyāḥ kṛṣikarā janāḥ ǀǀ471ǀǀ
sasyānāṁ kalamādīnāṁ vayo niścitya rūpataḥ ǀ
sākṣādanubhavāt bījaśaktyā ca sthalavaibhavāt ǀǀ472ǀǀ
śālyādīni supakvāni rakṣitāny anujīvibhiḥ ǀ
pradakṣiṇādibhir mānyānīha prāṇihitāni ca ǀǀ473ǀǀ
yadā tu dhānyakaṇikārāśau pakve yathākramam ǀ
tat tadvayaḥ kālayogāt tadā na jalasevanam ǀǀ474ǀǀ
kāryam ity evam ādiṣṭaṁ kṛṣitattvavidāṁ varaiḥ ǀ
śālikṣetreṣu sarveṣu kathito 'yaṁ kramo budhaiḥ ǀǀ475ǀǀ
tasmāt tadvrīkaṇikāḥ pakvakāle kṛṣīvalāḥ ǀ
atandrā rakṣeyur etāḥ doṣāt vividharūpataḥ ǀǀ476ǀǀ
evaṁ tadvrīhikaṇikāgaṇe pakve dalāgrake ǀ
vrīhikāṇḍavaliḥ sāgrā patati iha bhuvau tadā ǀǀ477ǀǀ
evaṁ tāṁ patitāṁ dṛṣṭvā dinaviṁśatikāvadhi ǀ
nirīkṣya pālayet kṣetre bhṛtyair vā sasyam eva vā ǀǀ478ǀǀ
atha saṁprāptakāle tu daṇḍānāṁ vrīhisantateḥ ǀ
supakvānāṁ svarṇavarṇacchuritānām api kvacit ǀǀ479ǀǀ
taddaṇḍakartanaṁ kāryaṁ śaṁkulādyaiḥ kṛṣīvalaiḥ ǀ
anujīvikavargaiś ca bhṛtyair anyais tathā narāḥ ǀǀ480ǀǀ
kartayeyuḥ daṇḍakāṇḍāvaliṁ kalamasantateḥ ǀ
ekadā vā pratyahaṁ vā mithaḥ sāhyakriyāparaiḥ ǀǀ481ǀǀ
kartanaṁ phaladaṁ prāhuḥ yathā śālyādidaṇḍakāḥ ǀ

nāśitā vā na naṣṭā vā vṛṣṭyādais taskarais tu vā ||482||
tathā tatkartanaṁ kāryaṁ sasyānāṁ vividhātmanām |
kartitavrīhidaṇḍānāṁ sāgrāṇaṁ tu kṛṣīvalaiḥ ||483||
sthāpanaṁ khalabhūmiṣu bhṛtyair anucarādibhiḥ |
trayaṁ vā paṁcakaṁ tatra dinānāṁ khalabhūmiṣu ||484||
sthāpanaṁ rakṣaṇaṁ śastaṁ mahāphalam udīritam |
ślathāgrāṇāṁ supakvānāṁ daṇḍānāṁ vrīhisantateḥ ||485||
khalabhūmyāṁ mardanena dhānyānāṁ kaṇikāvaliḥ |
patitā rājate ratnaguṁbhāvalir ivāmalā ||486||
punaś ca vrīhidaṇḍāgre mardite sati yatnataḥ |
vṛṣāṇāṁ mahiṣāṇāṁ ca pādaghaṭṭanakais tu vā ||487||
avaśiṣṭo dhānyakaṇaguṁbhaś cādhaḥ pataty adhaḥ |
evaṁ saṁmardanaṁ kṛtvā vrīhidaṇḍāvaleḥ khale ||488||
puṁjīkuryur dhānyakaṇān ghanān sārāṇ pṛthakkṛtān |
ghanahīnān peśalāṁś ca śūrpavātādivījanaiḥ ||489||
pṛthakkṛtya prayatnena puṁjīkuryuḥ kṛṣīvalāḥ |
ātape śoṣayitvātha nirmalān kārayet budhaḥ ||490||
devānām atha rājñāṁ ca bhāgam ekaṁ yathākramam |
srotriyāyāpi viduṣe bhāgam ekaṁ yathākramam ||491||
datvā dānaṁ bhṛtyavargapoṣaṇaṁ vā kvacit sthale |
avaśiṣṭaṁ dhānyarāśiṁ sthāpayet svaniketane ||492||
rakṣayec ca prayatnena kāṭhinyādiṣu yuktitaḥ |
mṛnmayeṣv api bhāṇḍeṣu sutapteṣv atha vā kvacit ||493||
dṛḍhakācair nirmiteṣu piṭhareṣv api vā kvacit |
dṛḍharajjukṛteṣv akteṣv api deśānusārataḥ ||494||
kvacit bhūmyāṁ dṛḍhasthalyām avaṭaṁ vā prakalpayet |
tatra sopānasaṁyukte vrīhidhānyādikaṁ kvacit ||495||
sthāpayet rakṣayet yuktyā jalataskarakīrakāt |
mūṣakād api duṣṭāc ca satvāt bhayavivarjitam ||496||

evaṁ tu kalamavrīhidhānyānāṁ rakṣaṇaṁ param |
bījānām api caiteṣām ātape śoṣitātmanām ||497||
rakṣaṇaṁ caivam ādiṣṭam uttamais tu kr̥ṣīvalaiḥ |
evaṁ nīte dhānyarāśau svagr̥hāṇi kr̥ṣīvalaiḥ ||498||
sasyavrīhyādidaṇḍānām ātape śoṣitātmanām |
gopādamarditānāṁ ca rāśiṁ mr̥dulatantukām ||499||
puṁjīkr̥tya kṣudramānaṁ svagr̥he goṣṭhakepi ca |
sthāpayitvā viśeṣeṇa palālānāṁ ca rakṣaṇam ||500||
dhenūnāṁ vr̥ṣabhāṇāṁ ca mahiṣāṇāṁ tathā kvacit |
mahiṣīṇām ajānāṁ ca mr̥gāṇām api vā kvacit ||501||
palālapūlakair etaiḥ jīvanaṁ śāstraniścitam |
ataḥ sarvatra deśeṣu dhīmantas tu kr̥ṣīvalāḥ ||502||
dhānyādibhir mānavānāṁ prāṇarakṣaṇakarmani |
catuṣpātprāṇarakṣāsu sasyakāṇḍapalālakaiḥ ||503||
mahānandāya kalpante kr̥tayatnā vidhānataḥ ||504||
āryāṇām upadeśāc ca deśācārakramād api |
tathā svānubhavāt yuktyā kr̥ṣikarmaratā ratāḥ ||505||
brāhmaṇāḥ kṣatriyā vaiśyāḥ śūdrāḥ saṁkīrṇayonayāḥ |
ye cānye bhūsaṁpadas tu phalitaṁ dhānyarāśikam ||506||
sumuhūrte sulagne ca vinyasya svagr̥hottame |
pūjayitvā dhānyalakṣmīṁ dhanalakṣmīm tathaiva ca ||507||
gandhaiḥ puṣpais tathā śuddhadravyaiḥ śāstraviniścitaiḥ |
brāhmaṇān bhojayitvātha pradatvā dakṣiṇāṁ śubhām ||508||
yaśaḥ sukhaṁ ca paramam aiśvaryaṁ prāpya bhūtale |
tathā devānugrahaṁ ca suputrāś ca kalatrakāḥ |
pautrādibhiḥ sametāś ca hr̥ṣṭā rājanti nityaśaḥ ||509||
śālikṣetreṣu vividheṣv ataḥ prāptakr̥ṣikriyāḥ |
mānavā yuktito deśe svakīye ca vanāntare ||510||
grāmeṣv api ca satkṣetre sthalyām api tathā kvacit |

sārakṣetre nityajale dvitīyaṁ kṛṣikarma ca ||511||
kṛtvā tatphalam apy atra prāpnuvanti mahodayāḥ |
pratyabdam evaṁ sarvatra dvirāvṛttikṛṣikriyā ||512||
kṣetreṣu bahurūpeṣu śasyate phaladāyinī |
ajāśakṛdgośakṛtkalatādīnāṁ samarpaṇaiḥ ||513||
kṣetrāṇāṁ sāram utpādya dvitīyāvṛttiriṣyate |
kriyākriyā yogyapadā phaladā ca prakīrtitā ||514||
pratyahaṁ vīkṣaṇāt śālikṣetrāṇāṁ mānanād api |
kṛṣeś ca kramayogāc ca śāntātmānaḥ kṛṣīvalāḥ ||515||
dhenvādīnāṁ rakṣaṇāc ca prāpnuvanty adhikaṁ phalam |
tasmāt kṛṣikramavido mitho maitryādibhāṣaṇāḥ ||516||
dvitīyāvṛttikakṛṣikāryāya kṛtamānasāḥ |
pratyabdaṁ gorakṣaṇaṁ ca bhṛtyānām api rakṣaṇam ||517||
bījānāṁ rakṣaṇaṁ cāpi kulyādīnāṁ ca rakṣaṇam |
jalāśayataṭākādihradānām api rakṣaṇam ||518||
khanitraśaṁkulādīnāṁ rakṣaṇaṁ khalarakṣaṇam |
vṛtes tu rakṣaṇaṁ cāpi kṣetrāṇām api rakṣaṇam ||519||
mukhyaṁ dharmam iti jñātvā tathā kṛtvā prayatnataḥ |
deśācāravaśāc cāpi kṛṣikarma mahodayam |
mahānandāya kalpante trivargasukhasaṁpadaḥ ||520||

(14) atha āḍhakādivividhadravyaprāptyarthaṁ kṛṣikramādivarṇanam
kāśyapa upadiśati

āḍhakāṇāṁ ca māṣāṇāṁ caṇakānām api kvacit |
mudgānāṁ ca kulatthānāṁ tilānāṁ ca kvacit sthale ||521||
āvāpaś cāpi bījānāṁ marīcijīrakādinām |
kṛṣikarma ca kartavyaṁ vyaṁjanaṁ tair udīritam ||522||
annaṁ tu vividhaṁ loke dhānyotthais taṇḍulādibhiḥ |
kriyate prāṇarakṣārthaṁ tatkalpaḥ pākaśāstrakam ||523||

vīrasenena rājñā tu pākaśāstraṁ mahatkṛtam I
tatputreṇa nalenāpi gārgeṇa ca maharṣinā II524II
tatroktānnavyaṁjanādikramas tu vividho mataḥ I
tad eva vyaṁjanaṁ loke cānnādyānāṁ rucipradam II525II
māṣāḍhakamarīcyādyaiḥ lavaṇaiś ca guḍādibhiḥ I
saṁpādyate hi sarvatra puṣṭyārogyapradaṁ nṛṇām II526II
dhānyotpādanakāle ca caturais tu kṛṣīvalaiḥ I
vanyeṣv api ca bhāgeṣu api grāmyeṣu bhūtale II527II
āḍhakānāṁ ca māṣāṇāṁ mudgādīnāṁ ca kālataḥ I
sīrakṛṣṭesu bhāgeṣu svalpasikteṣu vāridaiḥ II528II
kulyānīrakaṇair vāpi bījāvāpaś ca niścitaḥ I
āḍhakeṣv api māṣeṣu mudgeṣv api tileṣv api II529II
caṇakeṣv api śāstreṣu bṛhallaghu ca niścitam I
jalasikte bhūmibhāge cottuṁge tatra tatra tu II530II
tilādīnāṁ tu bījānām āvāpaḥ kṛṣicoditaḥ I
catuḥpaṁcaṣasaṁkhyeṣu vāsareṣu gateṣu hi II531II
āḍhakādyāṁkurotpattiḥ dṛśyā sīrasthale tadā I
sā ca bhūsārayogena niścitā śāstrakovidaiḥ II532II
pratyakṣam ālokitā ca nānādeśeṣv api sthale I
vrīhikṣetraṁ tu bhūloke jalabhūmir atīritam II533II
māṣādikṣetrabhūmis tu tārabhūmir iti smṛtā I
tasmād āḍhakamāsādikṛṣikarma kṛṣīvalaiḥ II534II
tārakṣetreṣu kartavyam iti provāca bhārgavaḥ I
śiṣṭācārakramavaśāt saukaryād vā talasthite II535II
vrīhikṣetreṣu māṣādibījāvāpah kvacit sthale I
phalapradaś ca nirdiṣṭaḥ śāstre 'sminn api kāśyape II536II
yathādeśācāravaśāt sthalakālavaśāt tu vā I
jalabhūmikṣetrabhāge tārabhūkṣetrake 'pi vā II537II
sīrakṛṣṭe yathākālaṁ prāyaḥ samatale tathā I

bṛhadāḍhakabījāni māṣabījāni vā tathā ||538||
mudgabījāni caṇakatilagodhūmabījakān |
kulutthabījāni tathā kīrabījaṁ ca kodravam ||539||
bījāni yavarāṇāṁ ca marīcibījakāny api |
tūlaireraṇḍakabījāni jīrakāṇāṁ ca bījakam ||540||
bījāni sarṣapādīnāṁ tārakṣetrasthaleṣu vā |
vāpayeyur yathākālaṁ paṁktiśaḥ paṁktiśah kramāt ||541||
sītāśreṇīṣu vā nānārekhāśreṇīṣu vā punaḥ |
jalasiktāni bījāni kānicit kṛṣikovidāḥ |||542||
asiktāṁ ca jale tadvat kāṁś ca sadbījakāvalim |
vāpayeyur yathākālaṁ svadeśācāravaibhavāt ||543||
uptānām eva bhūbhāge dvividhe tatra tatra tu |
bījānām aṁkurotpattim ālokya kṛṣikāriṇaḥ ||544||
ekamāsena tu kiṁcin nyūnādhikyakramānvite |
kāle tale tu sarvatra pracurāṇi tṛṇāni hi ||545||
kṛṣīvalās tāni vīkṣya karaśaṁkulakādibhiḥ |
chitvā bhittvā samūlaṁ ca bahiruddhṛtya tatsthalāt ||546||
tṛṇādini tatas tatra jalasecanam ācaret |
bījāvāpānantaraṁ tu cāḍhakādisthale bhuvi ||547||
vṛṣṭyādhikyaṁ yadi tadā bījanāśo bhavet dhruvam |
ataḥ svalpaṁ jalakaṇāvalisecanam uttamam ||548||
kulyānīreṇa vā caitat kalpyam āhur munīśvarāḥ
tilagodhūmacaṇakamāṣamudgāḍhakavaleḥ ||549||
bījānāṁ nāśajanake saṁniraste tṛṇādike |
tadā śākhopaśākhānāṁ aṁkurādaṁkurāntaram ||550||
mūle puṣkalasāraṁ ca kalayet kalamādivat |
atas tṛṇānāṁ kandānāṁ śatrūṇāṁ nivahe hate ||551||
māṣādidaṇḍāśākhādivardhanaṁ kila niścitam |
tadānīm eva śobhādiḥ puṣpādiś ca manoharaḥ ||552||

puṣpodbhavānantaraṁ tu kaṇiśānāṁ ca santatiḥ I
kapiśeṣv āḍhakādīnām utpattiś ca vilokitā II553II
māsatrayeṇa kālena kvacit tadadhikena vā I
māṣāḍhakādicaṇakagodhūmotpattir īritā II554II
mūlasthale tu salilasekavaibhavato bhuvi I
antaḥsāro bhavaty eva kapiśānāmitīritaḥ II555II
evaṁ māṣāḍhakatilaprādurbhāvas tu bhūtale I
sarvatra daivayogena nirmitaḥ śāstradarśitaḥ I
antaḥsāre ghanībhūte māṣādiśreṇikā kramāt II556II
phalāya dr̥ṣṭaḥ sarvatra paṁktiśaḥ kaṇiśāṁcale I
tadā patrādipatanaṁ jīrṇatā patanaṁ tathā II557II
śiras talānāṁ namanaṁ nirdiṣṭaṁ kr̥ṣikovidaiḥ I
kālānuguṇam evaṁ tu jalasecanam ācaret II558II
tilānām api godhūmarāśīnāṁ tārabhūmike I
kaniśādipraveśaś ca kālabhedaś ca niścitaḥ II559II
jātibhedo rūpabhedaḥ kr̥ṣibhedas tathā kvacit I
vijñeyaḥ sthalabhedena kr̥ṣiyogyā tadarhakā II560II
āḍhakādighanībhūte kaṇiśānāṁ gaṇe tathā I
supakve kṣetrabhāgeṣu tatkrantanam udīritam II561II
bhr̥tyaiḥ vīrair rakṣakaiś ca phalakāle tu rakṣaṇam I
mahāphalāya bhavati dhānyādivad ihoditam I
chinnānāṁ truṭitānāṁ ca kaṇiśānāṁ khalasthale II562II
daṇḍādyair mardanaṁ proktam ātape śoṣaṇaṁ tathā I
āḍhakīdaṇḍakāṇḍādīn vr̥ṣāśvādicatuṣpadām II563II
bhakṣanārtham ihādiṣṭān rakṣeyuś ca kr̥ṣīvalāḥ I
puṁjīkr̥tya khalasthāne svagr̥he svocitasthale II564II
goṣṭhacatvārabhāgeṣu cāthavā sthalayogyakam I
āḍhakīdaṇḍakāṇḍādirakṣaṇaṁ ca sukhapradam II565II
vrīhidaṇḍapalalādīn āḍhakīdaṇḍakāṇḍakān I

māṣagodhūmacaṇakapatradaṇḍādikān api ||566||
catuṣpātjīvadān puṣṭipradān api ca sarvadhā |
sarvartuṣv api paśvāder ārogyasukhadāyakān ||567||
puṁjīkuryāc ca rakṣec ca kāryajñas tu kṛṣīvalaḥ |
kṣudragrāmakaṭugrevavihīnān tu kṛṣīvalaḥ ||568||
śūrpādivinyāsayogāt vātavījanatopi vā |
pecanāc cāmanād vāpi nirmalīkṛtarūpakān ||569||
godhūmatilamāṣādīn āḍhakīsarṣapādikān |
kīrān mudgān athānyāṁś ca puṁjīkuryāt khalasthale ||570||
svagrahaṁ cānayed etān ātape śoṣitān tathā |
nirmalīkṛtadehāṁś ca caṇakāḍhakamudgakān ||571|
vyaṁjanāya tathā sajjān bhāṇḍeṣu ca pṛthakpṛthak |
sthāpayed rakṣayet kāle sukhasiddhyai kṛṣīvalaḥ |
dhānyādīnāḍhakādīṁś ca marīcyādīn bhuvasthale ||572||
deśācāravaśād evaṁ yuktyā ca kṛṣikovidāḥ |
tattajjātisvarūpānuguṇaṁ ca sthalayogyakam ||573||
kālānuguṇamanyāṁś ca kṛṣilabhyān viśeṣataḥ |
saṁpādayeyuḥ grāmeṣu pureṣu nagareṣv api ||574||
tadantikasthale kṣetrabhāgeṣu vividheṣv api |
kṛṣyādānaṁ tu niyamāt mahāphalam udīritam ||575||
jalabhūmau tathā tārabhūmau vā kṛṣiyogyakāḥ |
pratideśaṁ pratigrāmaṁ kodravaṁ yavaraṁ tathā ||576||
yavaṁ ca vardhayet kāle tad dhānyam api puṣṭidam |
tatkāṇḍadaṇḍaḥ paśvādijīvānāṁ jīvanapradaḥ |
ataḥ sarveṣu deśeṣu kodravaṁ yavaraṁ yavam ||577||
puṣṭipradaṁ mahāsāraṁ vardhayitvā viśeṣataḥ |
agastyam ikṣudaṇḍāṁś ca sthalayogyaṁ ca vardhayet ||578||
mahāphalapradaṁ cekṣukṛṣikarma praśasyate |
gajānāṁ paramādharāṁ śarkarāguḍakāraṇām ||579||

ikṣudaṇḍaṁ dṛḍhaṁ kālavardhanaṁ jalabhūmiṣu |
atha vā tārabhūbhāge kvacit kṣetre viśeṣataḥ ||580||
dīrghāvaṭaśreṇikaṁ tu kṛtvā tatra nikhātayet |
kadalīnāṁ ca sarvatra puṇḍrekṣuṇām iha kṣitau ||581||
na bījakalanaṁ dṛṣṭaṁ partyakṣaṁ vāpi śāstrataḥ |
śaratkāle tathānyatra samaye deśacodite ||582||
kṛṣīvalair ikṣudaṇḍasthāpanaṁ cāvaṭasthale |
sthāpanaṁ salilāsekapūraṇaṁ ca kramoditam ||583||
gate daśadine kāle tanmūlādaṁkurodbhavaḥ |
ikṣudaṇḍāṁkuravaśāt dalādīnām athodgamaḥ ||584||
māsadvayena vā māsatrayeṇāpi kvacit sthale |
saṁpūrṇadalavṛddhis tu niścitā phaladāyinī ||585||
purusākṛtikaunnatyam adhikonnatir eva vā |
ikṣudaṇḍāvalīvṛddhiṁ jñātvā kṛṣivicakṣaṇaḥ ||586||
tanmūle cāvaṭe bhāge nityaṁ salilasecanam |
kalpayet yatnataḥ kāle rakṣayec ca viśeṣataḥ ||587||
vātādibhis tathā caiṣām ikṣūṇāṁ rakṣaṇaṁ param |
evaṁ prakāraḥ kathitaḥ kadalīnām iha kṣitau ||588||
kadalī phaladā śaśvat tajjātir vividhā matā |
deśācāravaśāt kālayogāt ca kṛṣikovidāḥ ||589||
kadalīm ikṣukāṇḍaṁ ca sarvartuṣv api vardhayet |
satataṁ phaladaṁ proktam etat sarvatra deśake ||590||
kvacit parvatasānau vā girimūrdhani vā kvacit |
kaṭhinasthalake bhūmisāraṁ jñātvā vicakṣaṇaḥ ||591||
kadalīm ikṣukāṇḍaṁ ca vardhayet sukhasiddhaye |
satataṁ phaladaṁ proktam etat sarvatra bhūmiṣu ||592||
na ṭūṣarasthale bhūmidoṣaduṣṭe tu vardhayet |
ataḥ sallakṣaṇopete bhūbhāge tu vicakṣaṇaḥ ||593||
sūpapradhānān caṇakāḍhakamudgādimān kvacit |

vardhayet rakṣayet tatra phalitān svagrahaṁ nayet ||594||
picuvṛkṣaṁ tūlavṛkṣaṁ tathā tacchreṇikām api |
kārpāsavastrasiddhyarthaṁ tārabhūmau kvacit sthale ||595||
tathoṣṇasthalake caiva vardhayet tu kṛṣīvalaḥ |
yat sukhapradam ādiṣṭaṁ paraṁparyopadeśataḥ ||596||
sākṣādvīkṣaṇato vāpi tat sarvaṁ vardhayet kṛtī |
khalabhūmitaṭe vāpi grāmānte grāmasīmani ||597||
ākrīḍāntaḥ sthale vāpi sthale samucite tathā |
jalāśayataṭaprānte cottuṁge grāmavāstuke ||598||
kārpāsavṛkṣavṛddhiś ca tathānyadvardhanaṁ kramāt |
śastam ity evam ādiṣṭaṁ śāstrajñair munipuṁgavaiḥ ||599||

iti kāśyapīyakṛṣisūktau dhānyādikṛṣi-
kramakathanaṁ nāma prathamo bhāgaḥ

atha śākādikṛṣikramakathana nāma
dvitīyo bhāgaḥ

kāśyapa upadiśati

kṛṣikāryakramavido loke sarvatra bhūmiṣu |
sallakṣaṇair bhāsurāsu nṛpājñāvartinaḥ sadā ||600||
śālikṣetre yathākālaṁ śālyādīn kalayet kramāt |
tārakṣetreṣu tadvac ca caṇakādīṁś ca vardhayet ||601||
etena bhojanopāyaḥ samagraḥ parikalpitaḥ |
śālitaṇḍulavargaiś ca saṁbhāradravyakais tathā ||602||
āḍhakādyaiś ca tatpūrtiḥ naiva dṛṣṭaḥ kṣamātale |
ataḥ śākādikān svādurasān tatra vivardhayet ||603||
jaṭikān rāśijaṭikān vallikā vanavallikāḥ |
paṭolikāṁś ca vārtākān śavakāṁś ca tathaiva ca ||604||
śākāṁś ca vividhān tadvat kūṣmāṇḍāṁś ca kalāṭakān |
kustuṁburūn kalāṭāṁś ca sūraṇān śākuṭāṁs tathā ||605||
haridrāś cāpi śuṇṭhīś ca sthalavanyaprabhedakān |

sthalībhavāṁś ca vividhān rasapūrṇān kṛṣīvalāḥ |
kṛṣikramāt vardhayeyuḥ bhojyakāryārthasiddhaye ||606||
etāni śākadravyāṇi mukhyānīti matir mama |
kvacit deśeṣu śākādiḥ jātirūpaprabhedataḥ ||607||
rasato varṇataś caiva bhinnaḥ saṁparikīrtitaḥ |
palāṇḍuṁ vardhayet drākṣānaladailādivardhanam ||608||
kṛṣīvalais tu kartavyaṁ tatra tatra sthalīṣv api |
tārabhūmitale vāpi jalabhūmitalepi vā ||609||
ṛtuyogyaṁ deśayogyaṁ pāraṁparyakramodbhavam |
rītiṁ ca vijñāya budhaḥ śākadravyaṁ vivardhayet ||610||
prathamaṁ taṇḍuladravyaṁ sūpadravyaṁ dvitīyakam |
śākadravyaṁ tṛtīyaṁ syāt kṛṣyādānakramodbhavam ||611||
ghṛtakṣīradadhidravyaṁ caturthaṁ ca samīritam |
etat caturvidhaṁ dravyaṁ bhojyakalpaḥ prakīrtitaḥ ||612||
sarveṣām iha devānām etat prītivivardhanam |
manuṣyāṇāṁ tu sarveṣāṁ jīvanasthairyakāraṇam ||613||
puṣṭidaṁ cārogyadaṁ ca budhyāyurvardhanaṁ tathā |
sṛṣṭikāle brahmaṇā tu sṛṣṭaḥ sarvatra bhūmiṣu ||614||
anena bhojyakalpena vasudhā tu sajīvanā |
brahmaṇaḥ sṛṣṭiṣāphalyam abhavat ca tadāditaḥ ||615||
sarveṣv api ca deśeṣu kṣetreṣu vividheṣv iha |
munīśvarāṇāṁ yamināṁ manuṣyāṇāṁ ca sarvataḥ ||616||
trivargaphaladaś cāyaṁ bhojyakalpaḥ prakīrtitaḥ |
tasmāt śākadravyakādikṛṣiś cāpi kṛṣīvalaiḥ ||617||
kalpanīyā viśeṣeṇa grāmeṣu nagareṣv api |
vane vanāntare vāpi sallakṣaṇasamanvite ||618||
janāvāsāntikasthāne jalādhare viśeṣataḥ |
śākādikṛṣikāryaṁ ca phaladaṁ ca viniścitam ||619||
vasantakāle grīṣme vā haimante vāpi vā kvacit |

vārtākavallījaṭikamarīcīṣavakādibhiḥ ||620||
sīrakṛṣṭaṁ kṣetratalam aṁcayed aṁkurārthibhiḥ |
vārtākādīni bījāni cātape śoṣitāni tu ||621||
gośakṛtkādibhiḥ sāramedure bhūtale kvacit |
aṁkurārthaṁ vinyaseyuḥ jalasekaṁ ca kovidāḥ ||622||
yathākramaṁ kārayeyuḥ palālācchādanaṁ tataḥ |
evaṁ dinatraye kāle galite tvaṁkurasthale ||623||
aṁkurāṇāṁ darśanaṁ tu kriyate vidhinā kila |
dinaviṁśatike kāle dṛḍham aṁkurajālakam ||624||
sīrakṛṣṭe sthale yogyaṁ cotkhātvā sthāpayet budhaḥ |
tadānīm eva mūle tu jalasekas tu jīvadaḥ ||625||
ālavālakramasthāne vārtākādīs tu vardhayet |
paṭolikāṁ vallikāṁ ca latārūpapravṛddhinīm ||626||
vardhayet vallikāvṛddhyai kāyamānasvarūpake |
grīṣmakāle jalakṣetre varṣākālaṁ vinā budhaḥ ||627||
kālāntareṣv api tathā śākadravyakṛṣiḥ śubhā |
śākuṭāṇāṁ tu kandaṁ ca sūranāṇāṁ ca kandakam ||628||
haridrākandakaṁ caivam āvāle vā prakalpayet |
sthale vā coṣṇabhūmyās tu kṛṣikarma phalapradam ||629||
itthaṁ latārūpakāṇāṁ kṛṣiś ca vividhā matā |
kūṣmāṇḍaṁ vanakūṣmāṇḍaṁ elānaladamukhyakān ||630||
vardhayet tārabhūbhāge nāgavallīm api kvacit |
kadalīmikṣudaṇḍaṁ ca nāgavallīṁ ca pūgikām ||631||
jalakṣetre tathodyāne cākrīḍe vardhayet budhaḥ |
rambhāṁ ca mocāṁ phaladāṁ panasaṁ likucaṁ tathā ||632||
rasālam āmraṁ ca tathā jaṁbūm api ca vardhayet |
nārikelatarūṁś cāpi pālayet kṛṣikarmavit ||633||
grahānāṁ niṣkuṭe vāpi cārāme vā kvacit sthale |
tārabhūmisthale vāpi jalabhūmitale kvacit ||634||

yathāsthalaprāpti loke brāhmaṇāḥ kṣatriyās tathā |
vaiśyāś ca śūdrāḥ saṁkīrṇajātijā api mānavāḥ ||635||
vyādhādayaś ca vīrāś ca yathāśakti svabhūmike |
dhānyakaṁ sūraṇaṁ vallīṁ kūṣmāṇḍaṁ ca paṭolikām ||636||
vardhayet śāstranirdiṣṭān jaṭilādīn yathākramam |
prathamaṁ bhūkarṣaṇaṁ tu dvitīyaṁ gośakṛtkaṇam ||637||
tatra kṣipec ca bhūsārakalanārtham idaṁ viduḥ |
tatra kṣetre bahumukham ālavālaprakalpanam ||638||
ālavālāṁcalasthāne vārtākādiniveśanam |
mālatīkundacāṁpeyakuṭajādisthaleṣv api ||639||
ālavālaṁ bahumukhaṁ kalpayet sthalayogyakam |
ālavālaṁcale bījasthāpanaṁ vāṁkurārpaṇam ||640||
deśācāravaśāt kālayogād evaṁ prakalpayet |
tato gate māsakāle vārtākādes tu mūlataḥ ||641||
tṛṇādivairisaṁghasya nirāsaś ca prakīrtitaḥ |
dinapaṁcakakālaṁ vā dinaṣaṭkam athāpi vā ||642||
vārtākādes tu mūleṣu cālavālamukhāt budhaḥ |
jalānāṁ secanaṁ kuryāt grīṣmakāle viśeṣataḥ ||643||
muhurmuhus toyadānaṁ tṛṇakoṣṭhanirāsakaḥ |
mahāphalāya nirdiṣṭam ṛṣibhiḥ śāstrakovidaiḥ ||644||
śākādipatrajāleṣu vārtākānāṁ tathā kvacit |
paṭolikānāṁ vallīnāṁ ṣavakānāṁ tathaiva ca ||645||
kūṣmāṇḍānāṁ kalāṭānāṁ sūraṇānāṁ tathaiva ca |
śākuṭānāṁ haridrāṇāṁ ārdrakānām api kvacit ||646||
patrajaleṣu kīṭādibhakṣaṇaṁ dṛśyate naraiḥ |
atas tat kīṭadoṣādinirāsārthaṁ kṛṣīvalāḥ ||647||
bhasmadhūlīkṣepaṇaṁ vā sudhāsalilasecanam |
kārayet kramataḥ kāle deśācāravaśāt tathā ||648||
bījānāṁ sthāpane cāpi tṛṇādikṣepaṇe tathā |

latādipālane cāpi puṣpāvirbhāvakālikam ||649||
yat karma niścitaṁ loke kriyate kāryakovidaiḥ |
svayaṁ nirīkṣya tat sarvaṁ paraṁparyopadeśataḥ ||650||
paṭolikāsu vārtākakūṣmāṇḍādiṣu bhūtale |
jaṁbīralikucādīnāmeladrākṣādikeṣv api ||651||
drākṣākharjurikādīnāṁ kṛṣyādiṣu subuddhayaḥ |
deśācāraṁ kramaprāptam ṛtuyogyaṁ kṛṣikriyām ||652||
kalayeyuḥ bhojyakhādyasvādyādisukhahevate |
teṣu kecit patrarūpāḥ surasāḥ parikīrtitāḥ ||653||
pare tu puṣparūpāḥ syur anye tu phalarūpikāḥ |
anye lavakarūpāḥ syur itare kandarūpakāḥ ||654||
puṣpodaye puṣpamadhye puṣpanāśo tathā kramāt |
kecid svādyāś ca peyāś ca kadalyāṁ tu viśeṣataḥ ||655||
kadalīkandakāṇḍaṁ ca puṣpaṁ lavakam eva ca |
phalaṁ ca kandakaṁ vāpi svādu rucyaṁ ca sarvadā ||656||
bhojyaṁ nirdiṣṭam etat tu munibhis tattvadarśibhiḥ |
paṭolikānāṁ lavakaṁ vārtākānāṁ tathaiva ca ||657||
śākānāṁ ṣavakānāṁ ca bālye svādutaro rasaḥ |
ato lavakam eteṣāṁ praśastaṁ parikīrtitam ||658||
panasānāṁ rasālānāṁ kūṣmāṇḍānām api kvacit |
lavakaṁ madhyavayasi rasaḥ svādutaro mataḥ ||659||
ikṣūṇāṁ nārikelānāṁ likucānāṁ tathaiva ca |
rasālapramukhādīnām antye vayasi bhūtale ||660||
temanādikriyāyogyarasas tatra nirūpitaḥ |
evaṁ svadeśācaraṇaṁ svasvajātikramād api ||661||
yad uktaṁ pākaśāstreṣu kāladravyādiyogataḥ |
svaśarīrasthapittādiguṇadoṣavaśād api ||662||
kṛṣikramavido loke paṭolādikṛṣiṁ tathā |
rakṣayeyuḥ gṛhe kṣetre cākrīḍe svīyabhūtale ||663||

kedaresv api vā vāpīhradādīnāṁ taṭeṣv api ।
jalāśayasyādhasthāt tu tanmukhe vā viśeṣataḥ ॥664॥
kadalīpūgapunnāganārikelāmravṛkṣakān ।
panasailādrākṣakādīn mālatyādīn viśeṣataḥ ॥665॥
vardhayed vārtākavallīkūṣmāṇḍān api yuktitaḥ ।
nānāśākān tathā peyān khādyān carvyān api kvacit ॥666॥
kṛṣyādānakramavaśāt rakṣayet pālayed api ।
deśe deśe ca sarvatra vasante grīṣmakepi ca ॥667॥
varṣākāle śaratkāle haimante cāpi vā kvacit ।
śiśirepi svakīyeṣu kṣetreṣu vividheṣv api ॥668॥
prakṛtvā cālavālādīn tattatkālānurūpataḥ ।
śākaṁ patraṁ tvacaṁ puṣpaṁ kandaṁ vā rasameduram ॥669॥
ārogyapuṣṭijanakaṁ kṛṣyādānavaśāt bhuvi ।
pālayeyuḥ kṛṣividaḥ grahnīyuḥ phalam uttamam ॥670॥
elālavaṁgārdrakādīn pūgān tāṁbūlavallikāḥ ।
ikṣukāṇḍāṁś ca kadalīranyāś ca vividhā latāḥ ॥671॥
kedārāṁcalabhāgeṣu tārabhūbhāgikepi vā ।
janopakārakānanyān pippalīpramukhān api ॥672॥
nānāvidhāś coṣadhīnāṁ latāś ca prāṇavardhakāḥ ।
vardhayeyur viśeṣeṇa tattatkṣetrānurūpataḥ ॥673॥
yathākālaṁ yathācāraṁ pūrvasūryupadeśataḥ ।
kṛṣisādhyaṁ prāṇivargasukhadaṁ yan nirūpitam ॥674॥
śāstreṣu munibhiḥ pūrvaiḥ tat sarvaṁ tu kṛṣīvalāḥ ।
yuktyā svasvakṣetrabhāge vardhayeyuḥ prayatnataḥ ॥675॥
svīyasvīyakuṭuṁbānāṁ rakṣārthaṁ brāhmaṇair api ।
prītyartham api devānāṁ kṣatriyair vaiśyakair api ॥676॥
śūdrair anyaiś ca bhūloke kalpyā kṛṣir udīritā ।
śāstreṣv api nibaddhā ca munibhis tattvadarśibhiḥ ॥677॥
sṛṣṭikartrā brahmaṇā tu kalpādau vividhāni tu ।

bījāni sṛṣṭāni bhuvi dhṛtāny atha viśeṣataḥ ||678||
bhūmidevyā prajāsṛṣṭiprārthanāpūraṇāya hi |
ekaṁ bījaṁ tu bahudhā rūpabhāk ca bhaviṣyati ||679||
bahudhā phaladaṁ nṛṇāṁ bhaviṣyanti na saṁśayaḥ |
iti bhūmyā viṣṇupatnyā pūrvam uktaṁ ca tat śrutam ||680||
kṛṣiśāstraṁ tu bhūmyoktaṁ nītaṁ tan munibhiḥ kṣitiṁ |
tac ca saṁrakṣitaṁ bhūpaiḥ lokarakṣādhuraṁdharaiḥ |
adhītaṁ tac ca kālena śūdrakais tu viśeṣataḥ ||681||
tadādi bahudhā śāstraṁ kṛṣyākhyaṁ bhūnivāsibhiḥ |
parjanyakālānuguṇaṁ vistṛtaṁ rakṣitaṁ ca tat ||682||
evaṁ kālena mahatā nānādeśeṣu mānavaiḥ |
pratyekaṁ kṛṣiśāstraṁ tu cādhītaṁ bhaktipūrvakam ||683||
kṛṣiś ca kalitā svīye kṣetre kedārakepi ca |
phalaṁ ca labdhaṁ saṁpūrṇaṁ poṣitā jīvarāśayaḥ ||684||
svaprajāś ca viśeṣeṇa devabhūpālasāhyataḥ |
ataḥ sarvatra deśeṣu kṛṣikarmaṇi bhūmipaiḥ |
uttamaṁ sāhyam ādiṣṭaṁ munibhiḥ kathitaṁ purā ||685||
vinā rājñāṁ tu sāhyena kṛṣir bhūmaṇḍale tathā |
deśe vā nagare grāme na saṁpūrṇaphaleritā ||686||
pratyakṣataś ca dṛṣṭaṁ ca prajābhir bhṛtyakair api |
ataḥ sarvatra bhūpālaiḥ kṛṣisāhyaṁ mahāphalam ||687||
kartavyaṁ maṇḍalānāṁ ca rāṣṭrāṇāṁ kṣemasiddhaye |
svātmakṣetrādivrḍḍhyarthaṁ yaśase ca parāya ca ||688||
devānāṁ prītaye cāpi svadharmasthairyahetave |
viprebhyaḥ śrotriyebhyaś ca nānāgotrebhya eva ca ||689||
bhūdānaṁ gopradānaṁ ca kulyākhananam eva ca |
kṛṣikarmaṇyaśaktānāṁ viprāṇāṁ ca viśām api ||690||
śūdrāṇām api cānyeṣāṁ hitāyānekarūpakam |
kṛṣisāhyaṁ tu kartavyaṁ rājadharmānusārataḥ ||691||

vāpīkūpataṭākānāṁ hradānāṁ khananaṁ tathā |
jalāśayānāṁ ca tathā ghaṭīyantraprakalpanam ||692||
sasyavidhvaṁsakānāṁ tu taskarāṇām anekadhā |
daṇḍaś ca śuddhasatvānāṁ yaśase dharmasiddhaye ||693||
evaṁ bahuvidhaṁ sāhyaṁ kr̥ṣikarmaṇi niścitam |
tat sarvaṁ bhūmipālais tu kartavyaṁ munibhāṣitam ||694||
yasmin grāme nagaryāṁ vā deśe vātha kvacit sthale |
kr̥ṣikarmaṇyaśaktānāṁ prajānāṁ rakṣaṇāya ca ||695||
janabāhulyanagararājadhānyādivāstuke |
upakārāya vā bhūpāḥ sārakṣetrāṇy anekadhā ||696||
kṣetrāṇāṁ maṇḍalaṁ vāpi dvividhānām mahītale |
svīkr̥tya kr̥ṣikāryārthaṁ yathākālaṁ yathākramam ||697||
niyojayeyuḥ svān bhr̥tyān prajāś cāpi viśeṣataḥ |
saṁkīrṇajātijān riktān śūdrāṁś cāpi kr̥ṣisthale ||698||
yojayitvā tatra tatra dhānyādīn vividhān api |
āḍhakādīn api tathā śākādīṁś ca kvacit sthale ||699||
mallicaṁpakamukhyāni kusumāny api yuktitaḥ |
amitaṁ vardhayeyuḥ svaprajākṣemābhivr̥ddhaye ||700||
śākādidravyarāśiṁ ca dhānyādīnāṁ ca rāśikam |
āḍhakīcaṇakādināṁ rāśiṁ ca tilarāśikam ||701||
māṣasarṣaparāśiṁ ca tathā drākṣādipuṁjakam |
vividhānāṁ phalānāṁ ca puṁjaṁ ca samayodbhavam ||702||
sajjīkr̥tya narāṇāṁ hi sukhāya sthāpayen nr̥paḥ |
āpaṇasthānake vāpi tadarthaṁ nirmite sthale ||703||
sacatvare sāṁgame vā nirāvaraṇakepi vā |
niṣadyāsu tathā śālānirmāṇeṣv api vā kvacit ||704||
sarvaṁ dravyaṁ ca pūrvoktaṁ yac cānyat sukhadāyakam |
kaṁbalaṁ vastrajālaṁ ca dadhikṣīrādikaṁ tathā ||705||
bhojyadravyaṁ ca vividhaṁ carvyaṁ khādyaṁ ca yad bhuvi |

kṛṣyādānavaśāt prāpyaṁ guḍaṁ tailaṁ hitapradam ||706||
yac cānyad dravyam atroktaṁ tat sarvaṁ bhūmivallabhaḥ |
grāme pure ca nagare rājadhānyāṁ viśeṣataḥ ||707||
niṣadyādisthānakeṣu cānīyāt tu kriyārhakam |
sthāpayitvā viśeṣeṇa vaiśyān vāṇijyakovidān ||708||
krayārthaṁ vikrayārthaṁ ca yojayet dhanikān nṛpaḥ |
vasantādiṣu māseṣu yat yat prāpyaṁ kṛṣikramāt ||709||
yantrādibhiś ca yat prāpyaṁ yat prāpyaṁ tantuvāyakān |
tailakārāc ca yat prāpyaṁ tat sarvaṁ pṛthivīpatiḥ ||710||
mālākārāc ca yat prāpyaṁ gandhadravyādikaṁ bhuvi |
saṁpādayitvā tat sarvaṁ prajānāṁ sukhasiddhaye ||711||
grāme pure ca nagare rājadhānyāṁ viśeṣataḥ |
āpaṇādiṣu śālāsu cāṁgane vividhepi ca ||712||
sthāpayet rakṣayet rājā vaiśyān tatra niyojitān |
prasthaṁ ca kuṁjaṁ droṇaṁ ca nāḍikaṁ kṣudram eva ca ||713|
dhānyāḍhakādidravyāṇāṁ kalanārthaṁ prakalpayet |
ayaḥpaṭṭaiḥ sāradāruphalakābhir mahīpatiḥ ||714||
prasthādikaṁ kārayec ca deśe sarvatra yojayet |
tailāditolanārthaṁ ca palaṁ kāraṁ tathaiva ca ||715||
ayaḥkhaṇḍena klṛptaṁ ca sarvatra samamānakam |
tulyarūpaṁ tulyadaraṁ deśe tatra niyojayet ||716||
tatpramāṇaṁ ca vaiśyeṣu grāmyeṣv api janeṣu ca |
nivedayet bhṛtyavargaiḥ brāhmaṇādyair viśeṣataḥ ||717||
tulādaṇḍaṁ ca kāṁśyaṁ vā pittalaṁ vā viśeṣataḥ |
śākāditolanārthaṁ ca rājā sarvatra yojayet ||718||
amātyapramukhai rājā palaṁ prasthādikaṁ kramāt |
aṁkayitvā viśeṣeṇa deśe sarvatra yojayet ||719||
dhānyādīnāṁ kriyajñānaṁ paṇajñānaṁ tathā nṛpaḥ |
udghoṣayec ca sarvatra deśe deśe viśeṣataḥ ||720||

pratigrāmaṁ pratigrahaṁ prajāsaukhyāya yuktitaḥ ।
dhānyādidravyarāśīnāṁ tailādīnāṁ tathaiva ca ॥721॥
vastrādīnāṁ ca śākādidravyānāṁ ca kr̥ṣikrame ।
yat sāhyaṁ vividhaṁ śāstranirdiṣṭaṁ munibhiḥ purā ॥722॥
tat sarvaṁ pr̥thivīpālaḥ kalpayet kṣemasiddhaye ।
kr̥ṣiṁ vinā na bhūloke dvipadāṁ ca catuṣpadām ॥723॥
sukhādir na bhaved eva niścitaṁ śāstrapāragaiḥ ।
tasmāt devaprīṇanārthaṁ prajānāṁ rakṣaṇāya ca ॥724॥
kr̥ṣikarmaṇyuparataiḥ nr̥pādhīnā ca niścitā ।
purohitaiḥ brāhmaṇaiś ca sacivair vā viśeṣataḥ ॥725॥
mahīpatiḥ kṣatriyo vā prajārakṣādhuraṁdharaḥ ।
sarvakāryeṣūttamaṁ tu kr̥ṣikarma praśasyate ॥726॥
tat pālayet svadeśeṣu yatnato yuktitaḥ kramāt ।
evaṁ mahīpatir yuktyā grāme vā nagarepi vā ॥727॥
rājadhānyāṁ viśeṣeṇa niṣkuṭaṁ vanam eva vā ।
purasya ca bahir deśe madhye vā vr̥kṣavāṭikām ॥728॥
antaḥpurasthapramadālīlārthaṁ vā viśeṣataḥ ।
ārāmaṁ sthāpayet dhīnāṁ vanānte vr̥kṣavāṭikām ॥729॥
sārāṇāṁ candanādīnāṁ vr̥kṣāṇāṁ vardhanād api ।
grahanirmāṇasāhyaṁ ca prajānāṁ parikalpayet ॥730॥
prāsādodyānabhāgeṣu manojñān phalamedurān ।
vr̥kṣāṁs tu vividhān bhūpo vardhayeyur yathākramam ॥731॥
rambhāṁ ca mocāṁ jambīraṁ mātulaṁgāṁ ca pippalīm ।
anyāś coṣadhivallīś ca naḷadān api vardhayet ॥732॥
kuśān kāśāṁś ca vividhān tulasīr bilvakān api ।
kusumāni ca pūjārhāny anekāni ca bhūmipaḥ ॥733॥
kālalabhyāni gandhāḍhyān api sarvatra vardhayet ।
kuṭuṁbārogyadāyīni yuktyānyāni ca vardhayet ॥734॥
devodyānaṁ devabhūmāvutsavāya prakalpayet ।

grāmasyāntaḥ sthale vāpi bahir vā tat prakalpayet ||735||
yathāsthalavaśāt bhūpo deśācāravaśād api |
evaṁ prāsādabhūmyāṁ tu pramadāvanam uttamam ||736||
manojñavṛkṣaṣaṇḍāḍhyaṁ mālatyādivirājitam |
sthāpayed evam anyatra rājadhānyā bahiḥ sthale ||737||
tac ca bhūpaḥ sthāpayec ca prajānandavivardhanam |
evaṁ vane vanānte vā girisānutaṭe 'tha vā ||738||
mahāvanaṁ sthāpayec ca nānāvṛkṣasamujvalam |
nānāvṛkṣāvalībījarakṣaṇaṁ tatra nirdiśet ||739||
phalaiḥ puṣpaiś ca phalakādārukhaṇḍādibhir nṛpaḥ |
prajānāṁ gṛhanirmāṇasāhyaṁ tena viniścitam ||740||
tasmāt sarvatra deśeṣu bhūmirakṣādhuraṁdharaḥ |
viśālabhūmibhāgeṣu sīmābhittiyuteṣu vā ||741||
karaṁjakān naktamālān eraṁḍān bhūrjakān api |
arjunān sarjakān nīpān kadaṁbān api bhūruhān ||742||
śighrūn śamān maruvakān devadārudrumān api |
śrīparṇakān tindukāṁś ca vaṭān plakṣān tathā kvacit ||743||
bhadradārun api tathā priyaṁgūṁś ca kvacit sthale |
vibhītakānāmalakīs tathā haimavatīr api ||744||
karṇikārāś ca likucān niṁbavṛkṣān api kvacit |
aśokāṁś ca śirīṣāṁś ca dāḍimān api bhūmipaḥ ||745||
jambūvṛkṣān kapitthāṁś ca sarvato bhadrakān api |
cāṁpeyakān kurabakān varṇajān api karṇikān ||746||
tāpiṁchāṁś ca tamālāṁś ca sinduvāradrumān api |
tintriṇyākhyān mahāvṛkṣān tathā vaṁśān api kvacit 747||||
cūtān rasālān āmrāṁś ca badarīn vā viśeṣataḥ |
rathadrumān madhūkāṁś ca piluvṛkṣān tathaiva ca 748|
punnāgān pāribhadrāṁś ca kovidāradrumān api |
evaṁ nānājātirūpān vṛkṣān sārān yathākramam ||749||

sthāpayet pr̥thivīpālaḥ rakṣayet jalasecanaiḥ ।
davāgnivījanāt duṣṭasatvāt corād yathākramam ॥750॥
evaṁ mahāvanaṁ rakṣyaṁ nadītīreṣu vā kvacit ।
gāndhārakuntipāṁcālakāśmīrāvantibhūmiṣu ॥751॥
sindhunepālaniṣādhakośalāṁgādibhūmiṣu ca ।
ghūrjarāvantisaurāṣṭrapramukhādisthaleṣu ca ॥752॥
deśeṣu vividheṣv evaṁ sārabhumisthale nr̥paḥ ।
mahāvanaṁ sarvavr̥kṣaṣaṇḍojvalitadiṅmukham ॥753॥
taṭākakulyāvāpyādisamyutaṁ ca samantataḥ ।
kr̥ṣīvalaiḥ svaprajābhiḥ sthāpayitvā yathākramam ॥754॥
vīrair bhaṭair yodhavargaiḥ rakṣayet sukhasiddhaye ।
parvatasyāntikasthāne vanamadhye 'tha vā kvacit ॥755॥
khaniṁ ca nānālohānāṁ janayitrīṁ mahīpatiḥ ।
parīkṣya bhūgarbhanāḍīṁ rakṣayet tatra tatra hi ॥756॥
ayaḥkhaṇḍaṁ pittalāṁś ca suvarṇaṁ rajataṁ tathā ।
gairikān api yatnena saṁpādya kṣemasiddhaye ॥757॥
lohakāryapravīṇaiś ca lohakāraiś ca takṣakaiḥ ।
suvarṇakārair api ca grāmeṣu nagareṣv api ॥758॥
nānāyudhānāṁ jālāni kalpayitvā viśeṣataḥ ।
kr̥ṣiyogyānyāyudhāni datvā grāmyajanāya ca ॥759॥
puraprāsādadurgādirakṣārthaṁ ca mahīpatiḥ ।
nānāvidhapraharaṇajālaṁ yodheṣv api kvacit ॥760॥
senāsthāneṣu ca tathā sthāpayet pālayed api ।
jāṁbūnadaiś ca rajataiḥ khanijaiḥ bhūpatiḥ kramāt ॥761॥
samudravīcījanitaprāvalaiś cāpi mauktakaiḥ ।
nānābharaṇajālāni kalpayitvā viśeṣataḥ ॥762॥
devebhyaḥ prathamaṁ datvā brāhmaṇebhyas tataḥ kramāt ।
svayaṁ dhr̥tvā ca mahiṣīṁ prāsādapramadā api ॥763॥
tais tair ābharaṇaiḥ kāmam alaṁkr̥tya yathākramam ।

paramānandasaṁpanno rakṣayec ca mahīm imām ||764||
evaṁ bhūpas tatra tatra deśe sarvasukhāvaham |
saṁpādayet tu vividhadravyajālaṁ tu sarvataḥ ||765||
kalpabhedena vā kālakrameṇa ca mahīpatiḥ |
kṛṣikarmavikalpaṁ ca bījāvāpavikalpakam ||766||
śītoṣṇasthānabhedena vibhaktāṁ ca kṛṣikriyām |
gopracāravaśād vāpi vṛṣṭipātakramād api ||767||
jalādharavaśāc caivaṁ jñātvā bahuvidhodayam |
kṛṣikarma prayuṁjīta sārakṣetrānusārataḥ ||768||
mahīpāle prasanne tu sarvaṁ sidhyati bhūtale |
mahīpāle 'prasanne tu sarvaṁ naśyati niścitam ||769||
tasmāt rāṣṭrasubhikṣārthī prasannaḥ pṛthivīpatiḥ |
kṛṣīvalāt karadravyaṁ paṇaṁ vā dhānyam eva vā ||770||
yathākramaṁ ca svīkṛtya bhūtalaṁ sacarācaram |
svakuṭuṁbasamaṁ nyāyyamārgeṇa parirakṣayet ||771||
brāhmaṇāḥ kṣatriyā vaiśyāḥ śūdrāś cānyepi mānavāḥ |
kṛṣikriyāsāhyam ārāt nṛpāt svīkṛtya bhaktitaḥ ||772||
yathākālaṁ yathācāraṁ saṁpādya ca kṛṣiṁ bhuvi |
karadravyaṁ ca datvā te rāṣṭrakṣemābhivṛddhaye ||773||
devānāgantukān svīyakuṭuṁbaṁ ca yathākramam |
poṣayitvā viśeṣeṇa dharmo mārgo yathoditaḥ ||774||
tathā śaktyā rakṣayeyuḥ yaśomaṇḍitadiṅmukhāḥ |
evaṁ grāme janapade nagare vā viśeṣataḥ ||775||
parasparasnehabhājaḥ prāptabhūpālasāhyakāḥ |
varṇāśramakriyāśaktyaḥ lasantu kṣmātale bhṛśam ||776||
iti kāśyapīyakṛṣisūktau śākādikṛṣikramakathanaṁ nāma dvitīyo bhāgaḥ
atha bhojyābhojyakramakathanaṁ nāma tṛtīyo bhāgaḥ
kāśyapa upadiśati
nṛpājñayā coditās tu brāhmaṇāḥ kṣatriyā viśaḥ |

anye cāpi yathākālaṁ sārakṣetre viśeṣataḥ ||777||
kr̥tvā kr̥ṣiṁ bahuvidhām evaṁ nirdiṣṭarītitaḥ |
dhānyānāṁ vividhānāṁ ca rāśimāḍhakarāśikam |
marīcīsarṣapādīnāṁ rāśiṁ svagraham ānayet ||778||
ātape śoṣayitvā tān yathākālaṁ yathākramam |
sajjīkuryus tu pākāya pratyahaṁ devatuṣṭaye ||779||
sādhvībhiḥ pākakāryaṁ tu kartavyaṁ devaśāsanāt |
devānām atithīnāṁ ca satkāraḥ saphalo bhavet ||780||
tataḥ prajārakṣaṇaṁ ca niścitaṁ tu munīśvaraiḥ |
dharmavit brāhmaṇas tasmāt dravyāṇāṁ bahurūpiṇām ||781||
bhojyābhojyakramaṁ jñātvā pāke sādhvīṁ niyojayet |
pīlvannaṁ kodravānnaṁ ca na devabrāhmaṇārhakam ||782||
śuṁbhaṁ ca śvetavārtākaṁ laśunaṁ graṁjanaṁ tathā |
palāṇḍuṁ kavakaṁ caiva niryāsaṁ ca tathākramam ||783||
abhakṣyaṁ ca brāhmaṇānāṁ kusthalaprabhavāni ca |
br̥hatkūṣmāṇḍakaṁ vāpi pannāṁ śigruṁ ca pīlukam ||784||
varjayet brāhmaṇas tatra cāsthalaprabhavaṁ tathā |
pipīlikāmakṣikādyaiḥ kr̥mibhiś cāpi yat grahe ||785||
dūṣitaṁ tac cānnamukhyaṁ varjayet yatnato dvijaḥ |
sārameyaiḥ kukkutādyaiḥ dr̥ṣṭam annaṁ tathaiva ca ||786||
viśvāsahīnaṁ dattaṁ ca dūṣitaṁ tu nakhādibhiḥ |
tuṣatantvādiduṣṭaṁ ca dvijo 'nnaṁ parivarjayet ||787||
mārjārair dūṣitaṁ yac ca yad āghrātaṁ ca veśmani |
cucuṁdarīmūṣakādyaiḥ yac cānnādyaṁ hi dūṣitam ||788||
tat sarvaṁ varjayet yatnāt brāhmaṇas tu viśeṣataḥ |
annādyaṁ nīcaduṣṭaṁ tu tathā ghrātaṁ tu vā kvacit ||789||
spr̥ṣṭaṁ vā na tu bhuṁjīta śreyo 'rtho sarvataḥ kṣitau |
aśuddhasthaladattaṁ ca sarvaṁ tac ca vivarjayet ||790||
kavalād avaśiṣṭaṁ ca cāsyāt patitam eva ca |

abhakṣyam āhur munayaḥ yaśase kṣemasiddhaye ||791||
adattam atithibhiś ca devatābhyo viśeṣataḥ |
akāladattaṁ ca tathā yad annādyaṁ grahādiṣu ||792||
asthale ca tathā paktaṁ jalenākṣalitaṁ tathā |
annaṁ vā śākarāśiṁ vā bhakṣyaṁ vā khādyam eva ca ||793||
abhakṣyam eva nirdiṣṭaṁ brāhmaṇānāṁ viśeṣataḥ |
akulīnena paktaṁ ca malamūtrādidūṣitam ||794||
romādidūṣitaṁ yac ca svenādau bhakṣitaṁ tathā |
asnātapaktaṁ ca tathā varjayet bhojanādikam ||795||
hastadattaṁ cāpasavyakaradattaṁ ca yat tathā |
ayaḥkhaṇḍena dattaṁ vā dārukhaṇḍakaraṁḍakaiḥ ||796||
tathā paryuṣitaṅ cānnaṁ varjayet dvijasattamaḥ |
grahāntare tu yat paktamānītaṁ svagrahaṁ tataḥ ||797||
tac cābhojyam ihādiṣṭaṁ munibhiḥ śāstrapāragaiḥ |
dadhyannaṁ ca tilānnaṁ ca rātrau bhuṁjīta no dvijaḥ ||798||
candrikābhojanaṁ caiva pradīpena vivarjite |
sthale ca bhojanaṁ tadvat bhojanaṅ tamasi kramāt ||799||
nindyam etac ca nirdiṣṭaṁ munīndraiś ca surair api |
saphenaṁ ca jalaṁ tadvat nadyādeḥ prathamodakam ||800||
ākāśāt patitaṁ toyaṁ dhārāsalilam eva ca |
aspṛśyaiḥ spṛṣṭatoyaṁ ca yatnato varjayet budhaḥ ||801||
uṣṭrakṣīraṁ caikaśaphakṣīraṁ dvikhurajaṁ tathā |
kṣīraṁ salavaṇaṁ tadvat yuvatikṣiram eva ca ||802||
vatsaprahīnagokṣīraṁ vatsāntaravinisratam |
gokṣīraṁ ca tathā loke hy ajākṣīraṁ ca dūṣitam ||803||
dugdhaṁ ca tāmravinyastaṁ varjyaṁ nindyaṁ ca tat viduḥ |
atas tu śrotriyo vipraḥ dharmavit brāhmasiddhaye ||804||
bhojyābhojyakramaṁ jñātvā pākaṁ kuryāt yathāvidhiḥ |
kṣatriyānāṁ ca vaiśyānāṁ śūdrāṇām api bhūtale ||805||

noktaḥ sa eṣa niyamo munibhis tattvadarśibhiḥ I
māṁsabhojanam apy eṣāṁ na doṣāyeti niścitam II806II
brāhmaṇāḥ kṣatriyā vaiśyāḥ śūdraś cānyepi bhūtale I
bhojyābhojyakramaṁ jñātvā deśācārapramānataḥ II807II
devatāyārpitaṁ pūrvaṁ bhuṁjantas tadanantaram I
dhūtapāpāś ca kāśante śraddhabhaktiguṇojvalāḥ II808II
tasmāt sarvatra deśeṣu brāhmaṇāḥ kṣatriyās tathā I
vaiśyāś ca śuddhadravyāṇi taṇḍulādīni nityaśaḥ II809II
śākadravyāṇy api tathā pāvanāni navāni ca I
jalaṁ dugdhaṁ dadhi ghṛtaṁ navanītam athāpi vā II810II
tailaṁ ca patravargaṁ ca pātrāṇi ca yathākramam I
svayaṁ vā svīyapatnībhiḥ sādhvībhiḥ svasvaveśmani II811II
gomayāliptabhūbhāge śuddhaṁ cullīṁ ca sendhanām I
pūjayitvā viśeṣeṇa bhūmiṁ vāsavam eva ca II812II
yamaṁ cāgniṁ ca varuṇaṁ brahmāṇaṁ ca viśeṣataḥ I
pākaṁ kṛtvā yathākālaṁ bhāṇḍaiś cāpi ghaṭādikaiḥ II813II
tataḥ śuddhasthale sthāpyaṁ ṣaṭrasādisamujvalam I
bhakṣyaṁ bhojyaṁ ca vividhaṁ vaṭakādikam eva ca II814II
pākaśāstroktamārgeṇa paktaṁ tailaghṛtādibhiḥ I
śarkarādyair guḍair nānāvyaṁjanādibhir anvitam II815II
temanādyaiś ca vividhaiḥ sopakārair viśeṣataḥ I
paktaṁ śraddhaprītipūrvaṁ yathākālaṁ yathākramam II816II
sthāpayeyur devatārcasthale veśmani vā punaḥ I
piṭakādiṣu bhāṇḍeṣu pūrayeyur yathārcitam II817II
nivedanāya devānāṁ tad etat śāstracoditam I
iti kāśyapīyakṛṣisūktau bhojyābhojyakramakathanaṁ II818II
nāma tṛtīyo bhāgaḥ

atha vividhahavyanivedamakramakathanaṁ nāma
caturtho bhāgaḥ

kāśyapa upadiśati

pākakriyā tu dvividhā nityanaimittikakramāt ।
nityaṁ tu nityaṁ deveśadṛṣṭipātādiyogyakam ॥819॥
naimittikaṁ ca tadvat syāt pātrabhedas tu tatra hi ।
devārhaṁ bhojyajālaṁ tu havyam ity abhidhīyate ॥820॥
idam evātithīnāṁ ca prajānāṁ śāstracoditam ।
bhojyam ity evam ādiṣṭaṁ kavyaṁ pitṛniveditam ॥821॥
tat tu svabandhubhir bhojyaṁ śāstrajñaiḥ samudīritam ।
deveśarūpaṁ vividham ādiṣṭaṁ pūjanārhakam ॥822॥
śivaliṁgasvarūpaṁ ca śālagrāmaśilātmakam ।
dhātur viṣṇoś cāvatārabhedaṁ jñātvā dvijottamaḥ ॥823॥
kṣatriyo vā vaiśyakaś ca yathāśakti vidhānataḥ ।
sāndhyaṁ karma tu nirvartya devapūjāṁ prakalpayet ॥824॥
rājopacāravat tat tu sakalaṁ śāstracoditam ।
gaṇḍūṣācamanādyaṁ tat snānavastrādidhāraṇam ॥825॥
bhojyāsanaṁ prakalpyātha śraddhābhaktisamanvitaḥ ।
sādhvībhiḥ paktam annādi vividhaṁ bhojyam uttamam ॥826॥
samarpayet devatṛptyai nānāvyaṁjanaśobhitam ।
guḍānnaṁ pāyasānnaṁ ca tilānnaṁ ca tathaiva ca ॥827॥
dadhyannam atha mudgānnaṁ apūpaṁ vaṭakaṁ tathā ।
godhūmavaṭakaṁ tadvat polikāṁ mājakām api ॥828॥
modakaṁ śarkarābandi drākṣailādisamanvitam ।
kuṁkumadravyasaṁyuktaṁ ghanaśarkarakānvitam ॥829॥
payogalakabandī vā vicitrānnam athāpi ca ।
yuktyā pākakramaprāptaṁ madhuraṁ kālayogyakam ॥830॥
māṣataṇḍulasamiśraṁ lavaṇādirasojvalam ।
ḍalaṁ mahāḍalam api pākaprāptaṁ ghṛtānvitam ॥831॥
tad eva dadhyanvitaṁ vā saghṛtaṁ ca saśarkaram ।
sudhārasasamaṁ paktaṁ ḍalaṁ ca vividhākṛti ॥832॥

śārkaraṁ jālepikaṁ vā mudgagodhūmasaṁyutam |
saghr̥taṁ māṣasaṁpannaṁ atha vā tat prakalpayet ||833||
madhuraṁ lavaṇaṁ tadvat kaṭuṁ tiktaṁ tathāmlakam |
kvacit tuvarakaṁ vāpi ṣaḍrasaṁ dravyam īritam ||834||
godhūmamāṣamudgānāṁ caṇakānāṁ ca piṣṭakaiḥ |
ḍalādibhakṣyaṁ vividhaṁ khādyaṁ peyaṁ tathā kvacit ||835||
carvyaṁ ca kramataḥ paktaṁ rasanārucidāyakam |
yuktyā paktaṁ yathākālaṁ taile vātha ghr̥tepi vā ||836||
annādyaṁ temanādyaṁ ca tathātirasamukhyakam |
bhakṣyaṁ ca vividhaṁ paktaṁ śraddhābhaktipuraḥsaram ||837||
śuddhadravyaiś ca saṁpaktaṁ niyamena mahānase |
etat sarvaṁ bhakṣyajālaṁ bhojyādi ca yathākramam ||838||
bhojyāsananiviṣṭāya viṣṇave 'mitatejase |
devārcanakramād bhaktyā cārpayitvā dvijottamaḥ ||839||
varivasyāṁ samāpyātha puṣpagandhānulepanaiḥ |
cāmaravyajanādīnāṁ vījanaiḥ stutibhis tathā ||840||
rājopacāravat sarvaṁ devārcanam udīritam |
kalpayitvā viśeṣeṇa nitye naimittike tathā ||841||
samaye bhaktivākyādyaiḥ praṇāmādibhir anvaham |
toṣayitvā jagannāthaṁ praṇatābhīṣṭadāyakam |842||
śuddhadravyaiḥ kr̥ṣiprāptaiḥ paktaṁ pratyaham ādarāt |
arpayitvātha kr̥ṣṇāya viṣṇave paramātmane ||843||
tatas tv atithivargāya svakuṭuṁbāya ca kramāt |
bhojyam annaṁ khādyavargam anyat paktaṁ mahānase ||844||
vividhaṁ ṣaḍrasopetam ātmarakṣaṇahetave |
arpayet bhojyapātreṣu kadalīpramukheṣu ca ||845||
bhojayitvā viśeṣeṇa prītipūrvaṁ dvijottamaḥ |
sacūrṇaṁ tāṁbūlaṁ ca datvā tebhyo yathākramam ||846||
evaṁ nr̥pājñayā loke dvijā dharmavidāṁ varāḥ |

devayajñaṁ bhūtayajñaṁ dvayam atra mahītale ||847
dharmyaṁ yaśasyam āyuṣyaṁ pālayet bhaktipūrvakam |
kuṭuṁbasya nṛpasyāpi rāṣṭrasya ca viśeṣataḥ ||848||
kṣemasidhyartham evaṁ hi kartavyam iti śāstrataḥ |
sarvalokahitārthāya kāśyapo munir abravīt ||849||

iti kāśyapamuniviracitāyāṁ kāśyapīyakṛṣisūktau vividhahavyanivedanakramakathanātmakaḥ caturtho bhāgaḥ

samāptā ceyaṁ kāśyapīyakṛṣisūktiḥ

gobrāhmaṇebhyaḥ śubham astu nityaṁ
lokāḥ samastāḥ sukhino bvavantu

iti śilāśāsanapariśodhakena paṇḍitena vijayarāghavācāryeṇa likhitam

Transcribed by pandit. V. Vijayaraghavacharya

Epigraphist
Tirupati. S. India
On 19th December 1930. A.D.

Notes on the Text

27b. Read: vārrāśijalapūreṣu
50b. Read: taṅkayāpi
70b. bhāga uttame for reasons of metre.
72a. Read: parighavattu or parighāyuktam
82b. Read: niścitā sukhadāyakā
92a. Read: bhūrjakam as in the verse 742.
155b. Read: mr̥ducchrayaṇam
159b. Read: ^{0}mr̥ducchrayaṇakāryataḥ
181b. Read: parānukūlaniratāḥ
190b. Read: śikṣyāḥ svavr̥ttayo anr̥śaṁsyam
218b. mauninā for reasons of metre.
232a. svakarmaniratā api for reasons of metre.
236b. Read: prītijanakaṁ śuddhadravyapradāyi
237a. lakṣmīvilāsakr̥t for reasons of metre.
240b. saṁkīrṇayā api for reasons of metre.
244a. Read: aṭajāṅgalabhūmiṣu
253a. Read: savatsāṁś ca
280b. poṣyāmyaham for reasons of metre instead of the grammatically correct poṣayāmyaham
334b. Read: prathamakarṣaṇād iha
365b. Read: palāśavrīhiś ca
417b. sapeṣṭake for reasons of metre.
447a. ta eva for reasons of metre.
451b. The scribe in his gloss proposed saṁnirasteṣu for saṁnirūddheṣu.
476a. Read: tadvrīhikaṇikāḥ.
477b. Read: vrīhikāṇḍavaliḥ sāgrā
555b. Read: antaḥsāro
557a. Read: dr̥ṣṭā
567a. Read: sarvathā
568b. Read: grāvavihīnām

572c. Read: bhuvaḥ sthale
603a. Read: dr̥ṣṭā
710a. Read: tantuvāyakāt
720a. Read: krayajñānaṁ
729b. Read: dhīmān
731b. Read: vardhayec ca yathākramam
741a. Read: bhūmirakṣādhuraṁdharaḥ
746b. Read: sarvatobhadrakān
749a. Read: rathadruma
752b. Read: Gūrjarāvanti^0 (Se)
756b. Read: rakṣayet tatra tatra hi
763b. prāsādapramadā api for reasons of metre.
804a. Read: viniḥsr̥tam
806a. Read: yathāvidhi
816b. Read: śarkarādyair
831a. Read: payoguḍakabandi (?) or payogaḍabandi (?) or payogolakabandi (?)
847a. Read: sacūrṇaṁ ca tāmbūlaṁ ca where the first ca must be inserted for the reasons of metre otherwise the verse is defective.

Glossary of selected terms unattested or insufficiently recorded in Sanskrit dictionaries

aṅkura-nyāsana m. the placing of sprouts 412.
aṅkura-vṛddhi f. the growth of sprouts 445.
aṅkurārpaṇa n. the placing of sprouts 640.
ajā-śakṛt n. goat-dung 430.
anujīvika m. a dependent (of a tenant cultivator) 480.
anujīvika-varga m. the group of the dependents (of a tenant cultivator) 480.
anna-śāla-sthāpana n. granary-building 200.
ayaḥ-khaṇḍa m. n. iron piece (used for making measures) 716, 757.
ayaḥ-paṭṭa m. an iron sheet [used for making measures] 714.
ājaka n. goat-dung 264.
āḍhaka-bhūmi f. pigeon-pea (Cajanus indicus Spreng.) field 329.
āḍhaka-sthalī f. pigeon-pea (Cajanus indicus Spreng.) field 339.
āpaṇa-sthānaka m. a market place 703.
āvāpa-vidhi m. the method of sowing 407.
ikṣu-kṛṣi-karma n. cultivation of sugar-cane 579.
ikṣu-daṇḍa-sthāpana n. the planting of scions of sugar-cane 583.
utkhanati digs out 625.
upakulyā f. a lateral canal 330.
uṣṇa-bhūmi f. a warm soil 629.
kaṭu? Unintelligible word. 568.
kaṭhina-sthalaka m. a hard ground 591.
kara-dravya n. tax paid in kind 770, 773.
kara-śaṅkula m. a small sickle used for weeding 546.
kartana n. harvest 482.
kartati harvests 481.
kartita mfn harvested 483. A similar form is attestable in the compound *sukartita* in a Buddhist Hybrid Sanskrit text (BHS D, 170); however the later is a PPP. from the causative of 2. *kṛt-* „twists, spins" (RVD, 23).
kartita-vrīhi-daṇḍa n. reaped stalks of rice 483.
karṣaṇa-yogyaka mfn. fit for ploughing (of a land) 337.
karṣita mfn. ploughed 406.
kala-vrīhi m. lit. 'soft rice', a variety of rice (Oryza sativa Linn.) 368.
kalama-vrīhi m. a variety of rice (Oryza sativa Linn.). 425. Normally, it is known as *kalama*.
kalāṭa m. Ceylon gamboge (?) (Garcinia morella Gaertn.) (SE 153). 386, 605, 646. Not quite clear.

kāṭhinya n. a stone vessel for storing grain 493.
kīra-bīja n. the seed of an unidentified plant. 539.
kulyā-khanana n. canal-digging 199.
kulyā-prakalpana n. canalization 128.
kūpa-nirmāṇa n. the constructuion of wells 161.
kṛṣi-karma-jña mfn. knowing agriculture (of a cultivator) 349.
kṛṣi-karmaṇyuparati f. the cessation of agriculture 725.
kṛṣi-karma-rata mfn. devoted to agriculture (of a cultivator) 505.
kṛṣi-karma-vikalpa m. the mode of agricultural technique 766.
kṛṣi-karma-vicakṣaṇa mfn. an expert in agriculture (of a cultivator) 410.
kṛṣi-karma-viśārada mfn. conversant with agriculture (of a cultivator) 388.
kṛṣi-kārin m. cultivator 544.
kṛṣi-kārya m. agriculture 196, 212, 215, 245.
kṛṣi-kārya-krama m. the order of agriculture 600.
kṛṣi-kārya-krama-vid mfn. knowing the order of agriculture (of a cultivator) 319, 600.
kṛṣi-kārya-jña mfn.knowing agriculture (of a cultivator) 247.
kṛṣi-kārya-rata mfn. devoted to agriculture (of a cultivator)170, 180.
kṛṣi-kārya-vid mfn. knowing agriculture (of a cultivator) 262, 269.
kṛṣi-kālānurūpa mfn. according to the season of cultivation 268.
kṛṣi-kovida m. an expert cultivator 126, 358, 417, 420. 589.
kṛṣi-krama m. the order of agriculture 606, 663.
kṛṣi-krama-vid mfn. knowing the order of agriculure (of a cultivator) 356.
kṛṣi-kriyā f. agriculture 247, 652.
kṛṣi-kriyā-sāhya n. help in agriculture 772.
kṛṣi-kriyā-sāhya-kārin mfn giving help in agriculture 195.
kṛṣi-kṣetra n. an agricultural field 172.
kṛṣi-kṣema m. n. the safety of agriculture 141.
kṛṣi-tattvavid mfn. knowing the essence of agriculture (of a cultivator) 475.
kṛṣi-paddhati-kovida m. an expert in agricultural manuals 110.
kṛṣi-prayojana n. the purpose of agriculture 173.
kṛṣi-pravṛtti f. livelihood from agriculture 242, 243.
kṛṣi-bheda m. a variety of cultivation 560.
kṛṣi-yukti f. agricultural expedience 208.
kṛṣi-yukti-pravartana n. employing agricultural expedience 206.
kṛṣi-yogya mfn. fit for cultivation (of soil) 55, 113.
kṛṣi-rakṣaṇa n. the guarding of the plough-land 220.
kṛṣi-labhya mfn. fit for agriculture (of soil) 574.
kṛṣi-vṛtti f. agriculture as an occupation 231.
kṛṣi-śāstra-viśārada m. an expert in agricultural science 47.
kṛṣi-sādhya mfn. to be attained by agriculture 674.
kṛṣi-sāphala n. success of agriculture 304.
kṛṣi-sāhya n. help in agriculture 197, 198, 687, 691.

kṛṣyādāna-krama-vaśāt on account of the order of making profit from agriculture 667.

kṛṣyādāna-vaśāt on account of making profit from agriculture 670.

kṛṣyārambha m. the start of ploughing 272.

kedārāñcala-bhāga m. a place at the border of field-beds 672.

kṣatra-bhūmi f. a soil possessing *kṣatriya* (=second class) quality 38.

kṣīra-pūrṇa mfn. full of milky juice (of paddy-ears) 463.

kṣīra-vṛddhi f. the increase of the milky juice (of paddy-ears) 466.

kṣudra-taluka m. a kind of agricultural implement 265. The element *taluka* has variously been translated as ' hoe' (Wojtilla 1985) or ' horn' (SE 87); I tried to explain it for

kṣudratalaka 'a roller or leveller used in rice-fields (Wojtilla 2003, 43); not any is sound.

kṣudra-rūpa-rakṣā-mṛttika-bhittikā f. small size wall of earth for the protection (of a rice-field) 331.

kṣetra-paṅkti-svarūpa n. the closeness to a line of fields 332.

khaḍgaka m. a long knife for cutting grain 265.

khala-bhūmi f. the land of the threshing-floor 95, 484.

khala-bhūmi-taṭa m. a slope of the threshing-floor 597.

khala-rakṣaṇa n. the protection of the threshing-floor 519.

khala-sthala m. the ground of the threshing-floor 370, 402, 570.

khala-sthāna n. threshing-floor 564.

khādira m. According to Meulenbeld it is an equivalent of *viṭkhadira* Acacia farnesiana Willd. or *arimeda* 'Acacia leucophloea Willd. 91.

kheṭa m. paring plough or a club for crushing clods of earth (Wojtilla 1985, 200) 414.

giri-prasravaṇārabdha mfn. Depending on mountain cascades (of a canal) 123.

guḍānna n. rice mixed with dry sugar 828.

godhūma-vaṭaka m. round cake made of wheat fried in butter or oil 829.

grāma-kārya-kāra m. village officer; in the title of chapter five

grāma-nyāya-sabhā f. the juridical council of a village 205.

grāma-rakṣaṇa-bhṛtya m. a hired labourer to guard a village 202.

grāma-vāstuka m. the built in area of a village. Cf. *vāstuka* 'a building site' (Sircar 1966, 367).

grāmāgra-gaṇya mfn. to be counted as the foremost of the village 190.

grāmya-jana m. village folk 759.

grāva-varjita mfn. free from stones (of soil) 126.

ghaṭi-yantra-prakalpana n. the fixing of a water-lifting device 692.

ghaṭi-yantra-prayoga m. the employment of a water-lifting device 171.

ghaṭi-yantra-sthala n. a firm land for a water-lifting device 165.

ghanavrīhi m. lit. 'compact rice'. A variety of rice (Oryza sativa Linn.). 365.

catur-māsa-phala-da mfn. ripening within four months (of rice) 459.

camana n. (?) Unintelligible form. It is the name of an operation, perhaps denotes the husking of pulses. 569.
churikā f. a short knife for cutting grain 265.
jala-kulyā f. a canal 337.
jala-kṣetra n. a wet field 410.
jala-poṣaṇa n. water supply 141.
jala-bhūmi-kṣetra-bhāga m. part of a field belonging to watery land 537.
jala-vyaya-sthāna-yuta mfn. endowed with a place where water comes out, i.e. endowed with a sluice. 90.
jala-sikta mfn. 1. sprinkled with water 410, 542. 2. irrigated with water 530.
jala-secana n. the sprinkling with water 547.
jala-srava m. a canal 334.
jalāśaya-sthāpana n. the building of a water-reservoir 200.
jātyagra-gaṇya mfn. to be counted as the foremost of the caste 190.
ḍala m. a kind of food offered to gods. Unexplained. 832, 833, 836.
taṭa-rakṣaṇa-bhṛtya m. a hired labourer to guard embankments 202.
taṇḍula-dravya n. rice ingredient 611.
tāra-kṣetra n. a table land 534, 601.
tāra-bhū f. a table land 534, 580.
tāra-bhū-kṣetra n. a field situated on table land 537.
tilānna n. rice mixed with sesame 828.
tulā-daṇḍa n. cross-bar of a balance 718.
tūlā-vṛkṣa m. cotton tree (Gossypium arboreum Linn.). 395.
tṛṇa-kāṣṭhotkṣepaṇa n. weeding 456. The exact meaning of *tṛṇakāṣṭha* is not clear. May it refer to taking out weeds from the roots?
tṛṇa-koṣṭha m. weed 446, 450. The meaning of *koṣṭha* is not clear.
tṛṇa-koṣṭha-nirāsaka m. weeding 644. The meaning of *koṣṭha* is not clear.
tṛṇādi-kṣepaṇa n. weeding 649.
toya-dāna n. irrigation 644.
tri-māsa-phala-da mfn. ripening within three months (of rice) 458.
dagārgala-pramāṇa-jña mfn. skilled in the standard of examining the soil in searching for water 47.
daṇḍa-kartana n. the cutting of the stalks (of grain) 480.
dadhyānna n. curd-rice 829.
dīrgha-śāli m. lit. long rice', a variety of rice (Oryza sativa Linn.) 362.
dravya-śāstra m. the science of implements 265.
drākṣā-vrīhi m. lit. 'vine rice', a variety of rice (Oryza sativa Linn.). 366.
dvitīya-vṛtti f. the second cultivation [in the year] 514.
dhana-go-dhānya-vṛddhi-da mfn. giving wealth, cattle and grain 54.
Dhana-Lakṣmī f. the goddess *Lakṣmī* in the form of riches 507.
dhānya-kaṇikā-rāśi m. the mass of ears of paddy 461, 474.
DhānyaLlakṣmī f. the goddess *Lakṣmī* in the form of grain 507.
nara-vāhyaka mfn. drawn by man (of a water-lifting-device) 168.

nava-dala-prāpti f. the appearance of new shoots 442.
nāga-vallī f. probably 'areca nut' (Piper bettle Linn.). 631.
pakka-kāla m. the time of ripening 476.
pañca-māsa-phala-da mfn. ripening within five months (of rice) 459.
pannā f. ? Unintelligible form. It stands on the list of plants the food prepared of which is prohibited for brāhmaṇas. 785.
payo-galaka-bandi m. a kind of globular sweetmeat made of milk and dry sugar 831. Unsatisfactorily explained. This form may stand for *payoguḍakabandi* (?) or *payogaḍabandi* (?) or *payogolakabandi* (?) Perhaps a kind of globular sweetmeat made of milk and dry sugar.
paraṁ-paropadeśa m. traditional advice 423.
palaśa-vrīhi m. a variety of rice (Oryza sativa Linn.) 365. The element *palaśa* is not clear.
palāla-cchādana n. a shed of straw (for protecting seeds) 623.
palāla-pūlaka m. a bundle of straw 502.
palāla-rāśi m. a heap of straw 400.
pāda-ghaṭṭanaka n. threshing by feet (of animals) 487.
pāna-ghaṭṭa m. a landing place for drinking water 94.
picu-vr̥kṣa m. cotton tree (Gossypium arboreum Linn.). 595.
piṭhara n. a vessel (made of strong glass for storing grain) 494.
pīta-toya mfn. watered (of soil) 445.
pīta-varṇa-vrīhi m. lit. 'yellowish rice', a variety of rice (Oryza sativa Linn.) 368.
pīlu-vr̥kṣa m. the tree Salvadora persica Linn. 396, 749.
puñjī-karoti piles up (stalks of grain or straw) 564, 570.
puṣkala-sāra n. plenty of manure 551.
puṣpodaya m. the beginning of the efflorescence 655.
pecana n. (?) 569. Unintelligible. It is the name of an operation, perhaps winnowing of grain. It is not quite unlikely that it stands for **pecchana-* the expected Middde-Indic form for Sanskrit *preṅkhana-* n. 'swinging' (Wojtilla 2003, 43). Ayachit (SE 104) equates it with *peṣana* 'pounding' with question mark, however, this meaning does not fit into the context. Or, has it anything to do with Telugu *pecu-* 'clean, peel, rind' (Brown 1980, 791)?
prathama-karṣaṇa n. the first ploughing 334.
prāptāṅkura mfn. having sprouts (of seed) 343.
phala-prada mfn. fruit-giving (of bean) 536.
phala-vrīhi m. lit. 'fructiferous rice', a variety of rice (Oryza sativa Linn.). 366.
phala-hīnatā f. the state of being without fruit 465.
bāhya-sāra mfn. having energy in the outside, i.e. in the upper layer (of soil) 40.
bīja-jīvada giving life to the seeds (of soil) 338.
bīja-vāpa m. the sowing of seeds 408.
bīja-vināśin mfn. seed-destroying (of soil) 31.
bīja-vinyāsa m. the sowing of seeds 438.
bīja-śakti f. the strength of the seed 472.

bīja-saṁrakṣana n. the preservation of the seeds 372, 388, 389, 394.
bīja-saṁgrahaṇa n. the collection of seeds 339.
bījāvāpa-kāla m. the time of sowing seeds 337.
bījāvāpa-vikalpa m. the art of sowing 766.
brāhmaṇa-kṣmā f. a soil possessing *brāhmaṇa* (=first class) quality 38.
bhasma-dhūli-kṣepana n. the scattering of ashes [mixed with] dusty soil 648.
bhū-karṣaṇa n. the ploughing of the field 267, 269, 411.
bhū-karṣaṇa-vidhi m. the performance of the ploughing of the field 276, 277.
bhū-garbha-nāḍī f. a tube in the womb of the Earth 756.
bhū-doṣākrānta-bhū f. soil suffering from terrain defects 132.
bhū-parīkṣā-krama-vid mfn. knowing the method of examining soil (of a cultivator) 46.
bhūpa-viśrāntika-sthala m. the resting place of the king 100.
bhūpāla-karma n. royal duty 215.
bhūmi-lakṣaṇa-paṇḍita m. expert in the characteristics of the soil 55.
bhūmi-sāra n. the fertility of the soil 591.
bhū-lakṣaṇa-vid mfn. knowing the quality of the soil (of a cultivator) 247.
bhū-sāra-yoga m. the using of manure in the field 532.
maṅgala-sūtra n. sacred thread worn by married women on their neck 16.
mardana n. threshing 486.
mardita mfn. threshed (of grain) 484.
mahā-ḍala m. a kind of food offered to gods. Unexplained. 832.
mājaka m. a kind of sweetmeat. Unexplained. 829.
mātṛkā f. a canal 109.
mukha-dvāra n. an opening gate (of a water reservoir) 164.
mudgānna n. rice mixed with *mudga* bean (Phaseolus mungo Linn.)
megha-vṛṣṭi f. rainfall from clouds 344.
rajju-jaṭila mfn. crested by ropes (of a plough) 256.
rāśi-jaṭikā f. Problematic. According to the copyist of the ms. it is identical with Tamil *kottavarai,* however, the latter can safely be identified with *jaṭikā* in the sense 'cluster bean' (Cayamopsis psoralioides) (Vacek). 384, 605.
latā-sthala n. a ground where creeping plants grow 418.
lava-sāraka m. Unintelligible. It has variously been translated as 'threshing apparatus' (AAI 37); by Ayachit as 'instrument for removing stubble' (SE 88); not any is sound.'Something that essence of which is the act of cutting'? 266.
lāṅgalādyarcana n. the worship of the plough and other [tools] 270.
loha-kārya-pravīṇa mfn. skilled in ironwork 758.
Vana-devī f. Goddess Forest 153.
vana-vallikā f. According to the copyist of the ms. it is identical with Tamil *kāttuppākal* 'prickly carolah' (Vacek). It can be a wild variety of *vallī* or *vallikā* a creeper. These two may denote quite different classes medical and other plants 384.
vallikā-vṛddhi f. the growth of creeping plants 627.

vāṇijya-vṛtti f. trade as an occupation 240.
vālaga n. Not clear. 264. It has been translated as 'compost' (AAI, 43). This meaning may fit into the context, however the form lacks explanation.
vṛṣa-vāhya mfn. drawn by oxen (of a plough) 262.
vaiśya-bhūmi f. a soil possessing *vaiśya* (third class) quality 38.
vrīhi-kaṇikā-gaṇa m. the group of ears of paddy 477.
vrīhi-kāṇḍa-vali f. a fold in the paddy stalks 477.
vrīhi-kṣetra n. a rice-field 533.
vrīhyupaci m. the ripening of the rice 334.
śaṅkula m. sickle 265, 480. Sanskrit lexicons record only the form *śaṅkulā* f. 'a kind of knife or lancet, pair of scissors' (Apte 1978, 1528). On the other hand *śaṅkula* as a Sanskrit loanword is know in Telugu in the sense 'chisel, nut cracker' (Brown 1980, 1247) which is a sickle-shaped tool indeed. The meaning 'sickle' can be secured on the basis of Modern Indo-Aryan forms such as *śaṅgiyā* 'a sickle with heavier blade in the dialect spoken in East Tirhut' (Grierson 1975, 73 and Kṛiṣikoś II, 316).
śarkarābandi m. A kind of globular sweetmeat made of sugar. 830. The element *bandi* is likely a borrowing of Telugu *banti* 'a ball' (Brown 1980, 865). Nene (SE 104) without English translation identifies it with *gujhiyā* or *karañji* (?). The first means either a sweet cake or bonbon made of condensed milk (Beskrovniy 1972, I, 464).
śāka-dravya n. a vegetable ingredient 610, 611.
śāka-dravya-kṛṣi f. cultivation of vegetable ingredients 628
śāka-bīja n. vegetable seed 403.
śākādi-kṛṣi-kārya n. vegetable husbandry etc. 619.
śākādi-sthānaka n. the cultivation place of vegetables and the like 172.
Śārngi-para m. the best archer, i.e. Viṣṇu 10.
śālā-nirmāṇa n. room under construction (?) 704.
śāli-kṣetra a rice-field 430, 439.
śāli-bījāṅkura-pālī f. a row of (*śāli*) rice seedlings 441.
śāli-bhū f. a rice-field 335.
śāli-bhūmi f. a rice-field 329, 334.
śāli-bhūmi-sthiti f. the situation of a rice-field 332.
śāli-saṁrakṣaṇa n. the protection of rice 457.
śāleya-bhūmi f.a rice-field 449.
śālyaṅkura m. rice sprouts 429.
śuka-vrīhi m. lit. 'parrot rice', a variety of rice (Oryza sativa Linn.). 365.
śuṣkita mfn. 'dried'. It would be a derivative form of a Buddhist Hybrid Sanskrit **śuṣkati* which is a denominative to Sanskrit *śuṣka.* In Buddhist Hybrid Sanskrit only the infinitive *śuṣkitum* can be attested (BHS D, 531–532) 421.
śūdra-bhūmi f. a soil possessing *śūdra* (=forth class) quality 39.
ṣaṇ-māsa-phala-da mfn. ripening within six months (of rice) 459.

ṣavaka m. Problematic. According to Nene (2002, 156) French tamarisk (Tamarix troupii Hole) The copyist of the ms. identifies it with Tamil *veṇṭai* 'ladyfinger, (Hibiscus esculentus = Abelmoshus esculentus Linn.) (Vacek) 385, 604, 620, 645, 657.
sa-kardama-jala mfn. full of muddy water (of soil) 439, 446.
sat-kula-saṁjāta mfn. arisen from a good stock (of cows) 306.
sa-piṣṭaka mfn. has the quality like flour 417.
sammara-vrīhi m. a variety of rice (Oriza sativa Linn.). 367. The element *sammara* is not clear.
sammardana n. threshing 488.
salila-secana n. irrigation 587.
sasya-kṣiti f. a rice-field 109.
sasya-kṣema m. n. the safety of rice 127.
sasya-bhūmi f. a rice-field 437.
sasya-ripu m. an enemy of the crop 447. It denotes various parasites and also bad weather.
sasya-vṛddhi-kṣaya-kṛt mfn. making destruction of the growth of rice 449.
sasyāṅkura m. rice sprout 433.
sāra-kṣetra n. a fertile soil 76, 88.
sāra-dāru-kṛta mfn. made of hard wood (of a plough) 256.
sāra-bhū f. a fertile soil 64, 68.
sāra-medura mfn. rich in manure (of soil) 622.
sāra-yuta mfn. endowed with manure (of soil), i.e. endowed with energy 40.
sita-vrīhi m. lit. 'white rice', a variety of rice (Oryza sativa Linn.). 368.
sītā-krama m. a series of furrows 414.
sītā-rekhā f. a row of furrows 415.
sītā-rekhā-kalpana n. furrowing 419.
sītā-śreṇī f. a row of furrows 542.
sīmā-bhitti-yuta mfn. endowed with boundary wall 741.
sīmā-sthala-vibhāga m. division of fields by artificially raised earth used as boundary 57.
sīra-kṛṣṭa mfn. ploughed (of land) 528, 538, 621.
su-karṣa m. good ploughing 261.
sūpa-dravya n. a sauce ingredient 611.
setu-saṁrakṣaṇa n. the protection of dams 220.
senāni-veśa-sthāna n. the place of the commander's house 100.
sthūla-kāya-vrīhi m. lit. 'big-bodied rice', a variety of rice (Oryza sativa Linn.). 365.
snāna-ghaṭṭa-avirodha m. the not disturbing of the bathing places 94.
svādu-vrīhi m. lit. 'sweet rice', a variety of rice (Oryza sativa Linn.). 366.

Appendix One

Rural society as seen in the Kāśyapīyakṛṣisūkti

We can only subscribe to Romila Thapar's opinion that 'The writing of a manual in Sanskrit on a particular subject was an indicator of its importance' (Thapar 2002, 306). As a matter of fact, the importance of agriculture in early Indian society may highly be overrated. Nene is certainly right in saying that 'Kashyapa has repeatedly stressed that agriculture is a virtuous profession and all people, regardless of their status in the society should practice agriculture. The idea, obviously, was to inculcate dignity of labor and ensure food security for all people' (SE 132–133).

It is well-known that the *Dharmaśāstras* hold different views about agriculture as a profession for *brāhmaṇas*. On the other hand, the Vedic literature maintains the essential role of agriculture in life. Verse 13 of the so-called gambler's song (RV X, 34) calls upon the gambler to abandon playing dice and encourages him to take to agriculture. Baudhāyana, the law-giver, has a different opinion, saying that 'The study of the Veda tends to the destruction of agriculture and (devotion to) agriculture tends to the loss (of the study) of the Veda. One who has the capacity (to look after both) may resort to both, but he who is unable (to look after both) should give up agriculture' (BauDS I, 5, 101 cf. Kane 1974 ,125). The classical law code, the MaS, contains strict prohibitory rules on this topic. With regard to the right livelihood of a *brāhmaṇa* MaS IV, 5 explicitly states that ' Gleaning and picking should be considered the "true"; what is received unasked is the "immortal"; almsfood that is begged is the "mortar"; and agriculture, tradition says, is the "fatal" (Olivelle 2005, 124). Furthermore we can find the following rules in MaS X, 83–84: 'A Brahmin, or even a Kṣatriya, who earns a living by the Vaiśya occupation, should try his best to avoid agriculture, which involves injury to living beings and dependence on others. People think that agriculture is something wholesome. Yet it is an occupation condemned by good people; the plow with an iron tip lacerates the ground as well as creatures living in it' (Olivelle 2005, 212). In the most important traditional Sanskrit lexicon, the Amarakośa, words related to agriculture stand in the chapter concerning the social group of *vaiśyas* (Ak. II, 9). According to R.Sh. Sharma, the tax-paying cultivator families of the Gupta period may have included both *vaiśyas* and *śūdras* (Sharma 1980, 259). Parāśara, the law-giver, permits a *brāhmaṇa* to be engaged in agriculture under certain circumstances and with certain conditions (Kane 1974, 126). VHā VII. 179 and 182 declares that agriculture is common to all social groups (*varṇas*) Verses 373–383 of the Atrismṛti discern ten kinds of *brāhmaṇas*. Verse 384 refers to so-called *vaiśya- brāhmaṇas* who deal with agriculture, animal husbandry and trade (Kane 1974, 126).

With regard to the question of whether *brāhmaṇas* may practice agriculture, the standpoint of the KṛP stands in diametrical opposition with MaS X, 83. Verses 2–3 of the KṛP declare: 'Even a Brāhmaṇa, who is versed in the four Vedas, who recites Śāstras and is wise, is possessed by Alakṣmi [Goddess of evil fortune] being humiliated by solicitation. By means of agriculture alone however, one ceases to be a suitor. By taking to agriculture, one can become a sovereign monarch in this world' (S.Ch. Banerji's transl.). A similar standpoint is taken in the Kṛṣiśāsana, a twentieth century Sanskrit treatise on agriculture composed by Daśarathaśāstrī, son of Nārāyaṇa, from the *Garga-gotra* of *brāhmaṇas* living in the village Śrīsukara i.e the modern Soroṅ in Uttar Pradesh (Wojtilla 2006, 38). The KāKS contains a series of distinct statements on the positive side of agriculture. 'The sages who have divine eyes praise agricultural work as the basis pf sacrifices and the life-giver of living beings' (verse 233). Agriculture eliminates dependency on others, Lakṣmī [Goddess of prosperity] manifests herself in agriculture, agriculture gives life for guests, gods and one's own family. Agriculture is praised because it gives various kinds of joy. Hence another occupation which has dependence on others does harm to meritorious acts and it is blamed by the people who see the truth' (verses 237–239a).

Prosperity of the subjects was one of the chief aims of kings in early India. There is an enormously great literature in Sanskrit on royal duties and also the on the non-performance of them which brings misfortune upon the realm. Among others, kings were expected to patronize agriculture. A perusal of the KāKS gives us a particular impression that the author/editor favours very strict royal supervision of agriculture and rural life.

The king – who bears the burden of protecting his subjects (*rājā prajārakṣādhuraṁdharaḥ*) – punishes the wrongdoers and establishes law exclusively. He should give nourishment to those subjects of good behaviour who are devoted to the system of classes and the four stages of life (*varṇāśramaratān*). Verse 24 states that 'In the various countries the king should divide this multi-faced Earth, which is lovely because she has lands suitable for agriculture.' By order the king should specify land for different purposes: for cultivation, for gardening, for water-reservoirs or left as forest (verses 56–57). The king is responsible for having villages, fortresses and towns built by architects (verse 60). 'At an auspicious moment, he should stations for herdsmen, hamlets of cowherds and abodes of cowherds – inhabited by a multitude of cows –, and a group of small villages accompanied by a water-reservoir and well adorned with gardens and surrounded by grain fields' (verses 62–63). He has to manage water-reservoir construction (verses 71–89); to afforest the bank of water-reservoirs (verses 90– 105); to construct canals (verses 106–108); to have wells dug by soldiers and cultivators and set with brick inside (verses 149–163).

The protection of the canals combined with water-reservoirs is a special duty of kings (verses 137–140). The king should procure water everywhere during the season and should strive to protect it (verses 178–179).

Verse 681 instructs us that the *kṛṣiśāstra* - meaning both the science of agriculture or the textbook of agriculture – was told by the Goddess Earth to the sages (*muni*) who brought it to Earth and then preserved by the kings who bear the burden of the protection of their subjects and in the course of time it was particularly studied by the *śūdras*. The sages (*muni*) advised the kings in the ancient times to direct the greatest help to agriculture (verse 685). 'Without the king's help agriculture is fruitless in the country (verse 686). He should give help to those who are incapable of doing agricultural work (verse 690).The help to agriculture (*kṛṣisāhyam)* is to be made according to the laws relating to kings (*rājadharmānusārataḥ*) (verse 691b). Verse 693 says that 'The punishment of thieves who damage the crops is for the glory of pure living beings and for the fulfilment of the law (*dharmasiddhaye*)'.

Measures affecting all rural economy and society emerge from verses 695–699: 'In villages, towns, provinces and other places, in order to give help to people who are incapable of agricultural work, or to settlements, such as overpopulated towns and especially the capital (*rājadhānya*), the kings should acquire fertile lands or a mass of twofold fields (i.e. wet and dry) and should use them for agricultural purposes in accordance with the season and (agricultural) methods. They should employ hired labourers and their subjects, particularly unemployed men belonging to the mixed castes (*saṁkīrṇajātijān*) and also *śūdras* in the agricultural location and should grow grain etc. and also pulses and vegetables in abundance.' Noteworthy here is the role of people of the mixed castes and the *śūdras*.

Verses 701–712 deal with the collection, storage and sale of agricultural products. It begins with the collection of vegetable stuffs such as grain, pulses, sesame, mustard, grapes and occasionally fruits. The king should store them and have them sold by rich *vaiśyas* - who are experts in trade. Anything which is obtained from agricultural work during the months beginning with the spring, things produced by mechanical implements, produced by weavers, oilmen, florists, perfumers etc., the king should collect for the sake of the fulfillment of the happiness of his subjects and should keep them somewhere and protect them in the village, city (*pure*), town (*nagare*) and especially in the capital, in the markets etc. (*āpaṇādiṣu*).' In comparison, the KA 2, 15, dealing with the royal magazine, lists all these articles except grapes. Our text strictly prescribes the units of weights and measures to be used when selling. KA economy is a store-reserve economy (Breloer 1934, 361–362), however, the KāKS represents rather market economy, strongly supported by the central government. This reminds us of the instruction given in the KN 5, 78–79 and AgP 239, 44.

Verse 726 is a summary of all that has been said earlier: 'The *kṣatriya* king, who bears the burden of the protection of his subjects, should protect agricultural work – which is praised as a superb activity among all occupations – with every effort in his country.'

This last sentence may also suggest that - according to our text - agriculture can be practiced by all castes. This basic idea finds a nuanced expression in a different part of our work.

On the topic of irrigation, a work of vital importance, verses 170–174 clearly state that the intelligent people who are engaged in cultivation, being *brāhmaṇas, kṣatriyas, vaiśyas, śūdras* and others, lead the water into a canal with the help of water-lifting devices.

Chapter five depicts the characteristic marks of cultivators and village workers in highly idealized form. Verses 180–185 read thus: ' Men who are devoted to agricultural work, living in villages, free from hypocrisy, seeking the welfare of each other, devotees of the cow and the earth-goddess, always truthful, content in giving comfort to others, whose mind is always peaceful, who have no enemy in the world, who aim at work for others, who have fondness towards the quadrupeds, who are experts in worldly affairs, skilful in counting, purified, who consider purity the ultimate aim, who devoted themselves to the constant and indispensable rites in proper time, who are doing good by receiving guests, who are free from quarreling and laziness, exempt from sensuality and anger, sharing in love towards each other, who are busy in the work to be accomplished – those best, truth-seeing people are to be assigned as guards of water-reservoirs, banks and canal.' *Brāhmaṇas, kṣatriyas, vaiśyas* and *śūdras* living from agriculture, or born in mixed caste, and men, who are hunters, who are fond of hunting, who are mentioned as cowherds, who are called goatherds and others who are living from agriculture, all these people in the world who are intended upon mutual happiness, who are initiated into the welfare of the world by the village headmen and the caste chiefs and by the king's orders, who do teaching for a living, who are watching the worldly affairs, merciful, ready for sacrifices for the protection of each life, who are eager to help everybody, those are called men' (verses 186–191). From among them *brāhmaṇas* should be employed by a king as chiefs of one or more villages (verse 194). 'In the case when *brāhmaṇas* are not available *kṣatriyas* or *vaiśyas* may be nominated there for the sake of the welfare of the subjects' (verse 207). Verses 209–211 speak of *śūdras* possessing special virtues whose employment is laudable. Verse 211b, however, definitely states that 'for the most part the *śūdras* as hired workers are known in the villages.' Verse 212a adds 'just they are to be employed in agricultural work and not the men of other castes.'

Signs of common work of village people can be read from verses 436–438: 'In village fields and in forest or other places, the ploughing of the rice-field at the same time by one, two, three, four or five cultivators, with the help of ten yokes drawn by twenty oxen, gives a great yield of rice. Simultaneous sowing by the village people of one accord and by the use of a number of yokes is said to be fruitful.'

Reaping is carried out by cultivators, accompanied by their dependants and hired labourers, 'who are mutually attentive to performing the task' (verse 480–481).

As we have seen, verse 681 definitely states that *kṛṣiśāstra* 'was particularly studied by the *śūdras.*' *Śūdras* began to be called peasants and agriculturists in early medieval texts (Sharma 1980, 261). Much earlier the KA informs us that the king should cause villages to be settled consisting mostly of *śūdra* agriculturists, with a minimum of one hundred families and a maximum of five hundred families (KA I, 2, 2).

As to the rules governing edible and inedible things, our text is quite liberal. Verses 808–809 say: '*Brāhmaṇas, kṣatriyas, vaiśyas,śūdras* and others also, having known the order of edible and inedible things according to the custom of the country, enjoy the food which is offered first to the deity and then they become sinless and adorned with the virtue of faith and devotion.' Although this instruction would primarily be valid for the whole kingdom, in the spirit of the KāKS it can be applied to a village or certain rural area where these different strata of people live and work together. Peace and welfare may come from unanimity. To highlight this atmosphere Kāśyapa refers to the sage Bhārgava calls people's attention to the birds who are always of one voice and one opinion (verses 217b-218). This ideal situation must be reasserted by the king's constant help and the activity of the village headmen, the caste chiefs and the juridical council of the village the members of which are nominated by the king, so declares verse 205.

In sum, the picture of rural society drawn by our text is rather *śāstric*, in other words highly typical. As a whole, it does not correspond to any particular rural area in any historical period. However, certain elements of it, such as the strong royal patronage or the rising tendency to regard *śūdras* as peasant cultivators and the indications of certain communal works, may refer to concrete situations in the early medieval times or later. At any rate, Yuan Chwang, the Chinese pilgrim, who travelled in India in A.D. 629–645, definitely states that in Northern India, the *śūdras* , as members of the fourth class of society, were the agriculturists and 'these who toil at cultivating the soil are industrious at sowing and reaping' (Watters 1973, I, 168).

Appendix Two

Technique of agriculture as seen in the Kāśyapīyakṛṣisūkti

Although long excerpts from the text were published in 1964 (AAI), they have escaped the attention of the experts since then. In the first place, the editors of AAI did not comment upon these passages. When Lallanji Gopal published his highly valuable paper on the technique of agriculture in early medieval India (Gopal 1963/64) he could not take them into account. Randhawa also satisfied himself with publishing selected passages merely to illustrate the development of agriculture in the period of the ninth century to eleventh century (Randhawa 1980, 482–490). In my paper, Rural Expansion in Early Medieval India (Wojtilla 1991, 163–169) I made a short review of basic trends in the period, however, allotted little room to the data of the KāKS. The commentaries by Nalini Sadhale and especially by Y. L. Nene have very much enriched our knowledge on basic aspects of agricultural work depicted in the treatise (SE 123–147), thus I shall very often refer to their findings. Both the bulky volumes 'Glimpses of the Agricultural Heritage of India' edited by Y. L. Nene (Nene 2007) and comprizing not less than eighty-four papers and the 'History of Agriculture in India (up to c. 1200 AD)' edited by Lalanji Gopal and V. C. Srivastava (Srivastava 2008) in the series History of Science, Philosophy and Culture in Indian Civilization, general editor D. P. Chattopadhyaya, consisting of forty-six papers lack a treatment of the KāKS. Lallanji Gopal, whose name is given as one of the editors, had died many years before this volume was published. Since he had informed me about the project before it was launched, I can say that the present volume only partially covers the original aims. It is only to be regretted that Prof. Gopal is only a nominal editor of it.

Being the lengthiest treatise on agriculture, KāKS is a rich mine of information on various aspects of agricultural technology. In the following I am going to exploit these data from which the outlines of a technical standard in agriculture, as it is seen in the KāKS, will emerge. I do all this in the form of comparison with the respective data of the KA and the KṛP, which represent landmarks of technology in agriculturein ancient and medieval India.

Agricultural meteorology

Nene rightly assumes that there is no discussion on agricultural meteorology and also no reference to famine in the text (SE 145). I think that the highly sophisticated system of irrigation, which must be ascribed to the special climatological conditions of the region where the core of our text comes from, explains the absence of these topics.

The knowledge of the soil

Verses 25– 59 contain a brief introduction into geology, beginning with the traditional concept of creation of Earth and entering into her geological features, especially from the point of view of agricultural production. Verses 32–45 give a colourful environmental study based on keen observation made by the sense organs. It reads thus:' Therefore the king who is eager to protect all of his objects should examine which land is free from pieces of bones and stone, which contains soft soil, which is very smooth, reddish and also dark coloured, free from wild grasses, husk and glass, (which is) strong and abundant in water, plain without holes, chasms and mounds, distinguished by the pleasant smell of *mallikā* (Jasminum sambac (L.) Ait.), *jāti* (Jasminum grandiflorum L.), *kuṭaja* (Wrightia tomentosa (Roxb.) Roem. & Schult.), *surāgandha* (Caesalpinia sappan L. ?) and which has a series of lotus, *kharjūra* (Phoenix dactylifera L.) and *tiniśa* (Ougeinia oojeinensis (Roxb.) Hochr.), which is not always flooded by water, but is saturated with water, which provides the seeds with rapid growth, which is pleasingly furrowed, which is smeared with the moisture of oxen's lip, and which is rich in offsprings of living beings, (which is) thornless, devoid of cow-dung etc., (which is) solid, compact, heavy in weight, which gives nourishment to herbs, which is classified into five groups: *brāhmaṇa*-land, *kṣatriya*-land, *vaiśya*-land, *śūdra*-land and land of mixed quality, which possesses clear water, which is endowed with energy in the interior and exterior also, which is fertile, which is a joy to the eye, (which is) free from anthills and so on, free from vicious animals, but home to useful animals and birds, which is not liable to winds and water-storms and wild fires, which contains inexhaustible water, which gives growth to pleasure gardens and orchards and to shady trees, which has little holes, which makes different seeds to grow again and again, which is soft to touch, always pleasant to cows and cattle, which has favourable qualities and only small defects, in which it is easy to reach water streams, and which is surrounded by threshing floors and which possesses lucky marks. The king having examined the land, should take possession of the best of the world.'Verses 50b-53 describe a soil test which, as Nene rightly puts it, is very similar to what is done today (SE 134):' The wise man, having dug up the land of some point with a spade or hatchet, looking at it he should examine it by measuring the soil carefully on a balance, by smelling and tasting it and then, putting it in a pot of water for about half an hour. He should test now the quality of the soil by the colour, the drop of water (turbidity?), by taste and by fluidity. Soil which has a uniform shade of colour, which is compact and smooth is regarded by the sages as excellent for gods, kings, *brāhmaṇas,* and *vaiśyas*. It furthers prosperity and growth.'

A separate chapter (verses 319–339) tackles the proper season for grains and the division of land. Taking a brief review of a series of regions in India,

Kāśyapa means that 'cultivation is very much laudable when it rains', however, 'cultivation is also blessed in such places which are irrigated by culverts connected with deep water-reservoirs and canals. The wise cultivator having examined the water-current coming from the clouds, from the deep water-reservoir and the quantity of water of the canal, should properly carry on cultivation keeping an eye on the output of the soil. The fields sitated everywhere on the banks of rivers, in villages, in the interior of forests, on the slopes of mountains, on high or low grounds are divided into two groups according to their form in this world. The rice-field is the first that is said to be the best by the sages.' And ' The raised field near the village habitat, near the forest or on the bank of a river is the second kind of fields. It is called *āḍhaka*-field. It needs only drops of water and grows *caṇaka* (chick-pea, Cicer arietinum L.)) and other pulses. For that canals and the like are not needed.' It is well in accordance with the advice found in the KA II, 24, 22.

Crop rotation

Although there is evidence for the rotation of crops in the early medieval period (Gopal 1963/64, 26), one looks for the same practice in our text in vain.

Fertilizing with Manure

Using manure was widespread both in agriculture and horticulture in the early medieval times (Gopal 1963/64, 104 etc.). In connection with the cultivation of *vārtāka* and the like verse 622 says: 'experts should sow its seeds sprinkled with water to sprout in a soil rich with manure of cow-dung. Having properly ploughed the field the cultivator should place there cow-dung, goat-dung and compost (264). By using thes three kinds of manure the fertility of the land can be raised to a degree that makes it possible to obtain a second harvest (513b-514a). Referring to cultivating grain crops verses 430–31 state that 'After the clay of the rice-field had been softened by ploughing it was fertilized with goat-dung, cow-dung and tendrils of creepers (*latāvratati*). Nene correctly identifies the latter with green fertilizer. He also calls attention to the absence of organic manures such as bones or flesh. He suggests that because Kāśyapa was a believer of Viṣṇu, he had an aversion to using dead animal parts as manure, and having been a *brāhmaṇa* 'was very strict about hygiene' (SE 145). I may add that this absence is rather puzzling in the case of the cultivation of certain vegetables and especially of planting trees, a subject-matter the author dwells upon at length in verses 741–756. As we stressed repeatedly, the treatise has gained its present form in a region where *Vaiṣṇavism* had an overwhelming position in religious life while the personality of Kāśyapa -does not matter- because he is only the nominal author of our text. Moreover, section three, which has really strict rules on edible and inedible things for *brāhmaṇas*, also says that 'The sages who see the truth have not declared this restriction for *kṣatriyas, vaiśyas* nor *śūdras* in this world. Even the eating of meat is not a sin for them – it is thus determined' (

805a-806). As Nene himself highlights, Kāśyapa encouraged all castes to practice agriculture SE 132–33), a circumstance that may permit the use of different kinds of manures by cultivators belongig to different castes.

Irrigation

Early medieval India saw an immense progress in irrigation techniques. The establishment of water-reservoirs and tanks, as well as the intensive canalization in many parts of India, can be traced in inscriptions and contemporary literary texts (cf. Gopal 1963/64, 1–14; Puri 1968, 383–390; Kulkarni,1990, 3–30; Thapar 2002 338–339). It is not by chance that Kāśyapa holds ancient Kosala the best country for rice cultivation (424). This positive opinion is based on its perfect irrigation system, in consequence of which the region is *adevamātr̥ka* i.e. not dependent on rain (Rām. II, 100, 45). The detailed description of the irrigation network supplied by lakes, rivers and wells which were filled by rainfal and strictly supervised by the king may remind us of the territory of the Pallava kingdom in the South, an area which witnessed rapid development of irrigation in the early medieval period (Adiceam 1966, 164–166; Myrdal-Runebjer 1988, 19–21). The chapters covering the method for establishing water-reservoirs (60–108) and the canalization (109–179) seem to reflect contempory reality.. The establishment of water-reservoirs is a royal duty. It must be located 'in the most suitable place on the west, north, east or south side of the villages and cities' in accordance with the situation of the land (70b-71). 'The reservoir should be deep, similar to a jar, resembling the shape of a bow, oblong, in some places circular, not shallow and well-protected by a strong buttress and causeway endowed with a high embankement and a road on it and the rain must find its way into it. Hence the king should establish the deep water-reservoir either in the vicinity of a hill or on a land table which is connected with a big lake or if it ison the plain the water-reservoir should becapable of holding abundant water. It should be established in firm land which is itself near to a perennial spring and which is free from sand and small stones and the like. The king should establish the deep water-reservoir which is fitted with flood-gates in such a manner that in the case of a full water-reservoir there is no danger to village or town, rampart, garden, fortress and threshing floor becaues of the running water overflowing or breaching [the dam]. The king should establish it in this manner everywhere and properly examine it in the country monthly, yearly, frequently. The king should appoint guards to the embankement. The guarding of the hundreds of canals and the like of the oblong-shaped water-reservoirs and tanks is approved to be carrying security, especially during the rainy season' (72–81). We learn further that 'The water-rervoir is called the main source of water. Hence the king, who bears the burden of the protection of his subjects, should establish and properly protect the water-reservoir with every effort. The king should establish one or two water-reservoirs according to the extent of the fertile land which is suitable to the village. He should have the water-reservoir constructed in such a manner that it be accompanied by a mountain rivulet or joined with a big lake or connected with a forest brook or a big river in different places of villages and towns. The water-reservoir must be endowed with an outfall. The king should have trees grown on the bank of the water-reservoirs' (86b-90). As to the importance of canals, verse 140 assumes 'the protection of the canals by all the kings is the highest law – so declare the sages who see the truth'. What is peculiar to the KāKS that is the enumeration of water-lifting devices. Verses 167–168 read thus:' There are various water-lifting devices. The one drawn by bullocks yoked with strong chains is the best. That drawn by an elephant with his trunk is the mediocre. The most inferior is the one drawn by man.'

Preservation of seeds, sowing and transplantation of seedlings

KāKS excels all of agricultural treaties in discussing the procuring and preserving seeds of grains, pulses, vegetables, creepers and flowers (350–351).Verses 358–400 enumerate an impressive list of plants of these four categories the seeds of which should beproperly collected 'by pure men and the best among the agricultural experts (356b-357a). A salient feature of the KāKS is its broad knowledge of rice varieties. Verse 360 says:' Keeping in view the different kinds of soil, the experts have declared twenty six varieties of rice here and there in the regions.' The list proper stands in verses 362–368. Nene justly remarks that the classification of rice types 'using planting method, duration, color of paddy husk, taste, shape and size of grain, extent of swelling after cooking, and medical properties deserve full appreciation and recognition' (SE 138).

Verses 408–423 explain the preparation of soil for sowing. Fields selected for rice cultivation are to be ploughed and then to be made smooth by levelling. These processes are followed by the plantating of seedlings. For planting bulbous plants and the like one should dig a pit. Banana and the creepers are cultivated either by the insertion of scions or bulbs or by growing seedlings. Verse 434 refers to the transplantation of seedling called *aṇkurajālānāṁ vinyāsaḥ*. Verses 439 – 442 graphically illustrate the result : 'In this manner, the rice seedlings transplanted in rows in a well-ploughed rice- field full of muddy water, in a village, town, forest or woodland, irrigated by a canal and provided with many outlets for draining away the surplus water, are made to take root by the Creator after the lapse of seven or ten days and there is a beautiful appearance of new leaves.' Verse 184 of the KṛP warns people from transplanting full grown paddy plants and verse 185 regulates the distance of each seedling at transplantation under different constellations.

Compared with the KA II, 24 11 where the different crops are divide into three categories according to the time of sowing, or with the advice given in verses 168–176, where the timing of sowing is determined by astronomical principles, our text speaks only of the due time that varies according to regional tradition. The careful preparation of different seeds, i.e soaking the seeds by night and drying in the heat by day , as we see in the KA II, 24, 24 or drying them up thoroughly in the sun and, purging the chaff and keeping them in a small pocket in the KṛP (157–158).Verses 542b-543 of the KāKS mentions seeds moist with water and good seeds not moistened in water. While Parāśara indulges in supertitious ideas that may hurt seeds (verses 159–164), verses 541b-542a of the KāKS have very much practical contents, saying that expert farmers should sow some seeds in furrows and some in rows divided by a variety of line marks.

Thinning out paddy seedlings and the removal of weeds from the paddy pictured in verses 186–192 of the Kṛṣiparāśara are only partially represented in our text. Verses 545–546 refer to uprooting wild grasses from the field, that had been ploughed twice and sown before its inundation, with the help of small sickle called *karaśaṅkula* (546) that had turned up in the field ploughed twice and sown, before its inundation. This tool may be identical with Hindi *khurpā* (Wojtilla 2003, 44).

Agricultural tools and operations

Plough and ploughing

The most important agricultural tool, the plough, is called *lāṅgala* (256, 261) *sīra* (verses 528, 532) and *hala* (256, 261). The description of the plough is incomplete. While in the early medieval period the number of distinct accessories of a plough may be five (Mās. V, 56–57 and KŚ I, 45–53), eight (KṛP, verses 112–117) or even nine (BPs.: Dharmaśāstrasaṁgraha, II, pp, 106–107), verse 256 in our text mentions the accessories only collectively (*sāṅga* i. e. 'together with the accessory parts'), speaks of the ropes used to harness the plough or to fix these accessories and makes an indirect reference to the pole (*īṣā*) made of hard wood. This laconic description does not speak of the material of the ploughshare. If a wooden plough is meant here, it represents a very simple implement, because the above mentioned sources record iron or iron-tipped ploughshares. Although we are told that the plough is drawn by bulls (*vṛṣavāhya*), their number is not specified. For instance verse 96 of the KṛP states that 'A plough is said to have eight bulls, six for ordinary use, four for the cruel and two for the cow-killers.' Kosambi mentions a plough drawn by a twelve-ox-team in modern times (Kosambi 1970, 89). Instead of an explicit statement we must be satisfied with a hint implied in verses 283–284: 'Having so requested the bulls and worshipped them, the work of a bull is laudable as far as he is not fatigued or exhausted. Affliction of their minds definitely causes fruitlessness. Therefore [the cultivator] should not employ fatigued bulls for agricultural work.' This attitude brings to mind the statement of verse 84 of the KṛP: 'Carry on cultivation in such a manner as does not toinflict pain on the draught-animals. Grains, obtained through the suffering of draught-animals, are condemned in all the rites.'

The way of ploughing is also peculiar in our text. Verse 263 reveals that the ploughing of a field must be made four or five times for six days or more. This procedure is likely intended for deep ploughing. The work appears inefficient, perhaps due to the simple plough implied here. One should keep in mind that other sources i.e. traditional lexicons in Sanskrit record fields ploughed two-times and maximum three-times. Otherwise deep ploughing can be achieved by lowering the ploughshare with the help of ropes (BPs.) or by applying an iron ploughshare and four or more bulls.

Harrow and harrowing

Verse 414ab assert that 'the levelling of fields by spades or hoes *(khanitra*) and the like and a paring plough (*kheṭa*) has been praised.' The term *kheṭa* seems to be a paring plough, however, it may also denote a club for crushing the clods of earth. It is a *deśi* word. (Wojtilla 1985, 200). Using spades or hoes for levelling is common: *viddhaka* (Kr̥P. verse 118) can be a big hoeing instrument (Ganguly 1931, 23) or a kind of rake. Another types of harrows, such as a ladder-shaped one called *madikā* (Kr̥P, verse 118) or *mayikā* (Kr̥P, verse 182) or *viddhaka,* a kind of big rake with twenty-one spikes (Kr̥P, verse 118) are missing from our text. According to verse 182 of the Kr̥P after sowing, the *mayikā* should to be applied there. In the absence of that, there is no even growth of the seeds.'

The tools and process of reaping

The inventary of tools for reaping and the account oof reaping in our text is far and away the best in Sanskrit texts bearing on agricultural work. Our text covers multiple reaping tools in addition to the traditional sickle (*lavitra*) (verse 266). Another types of reaping tools are sickle, a curved knife similar to what is used as a nut-cracker, called *śaṅkula* (verses 265 and 480), a 'long knife', *khaḍgaka* and a short knife, *churikā* (verse 265). Interestingly, these three words do not occur in any other Sanskrit text in this sense. The name *kṣudrataluka* (verse265) has been translated by Ayachit as 'small horn' without further explanation. This interpretation is unsound because the word *taluka* does not have this meaning. For the time being, the compound word must be regarded as unexplained. The main instrument for reaping in our text is *śaṅkula.*

Verses 474–482 report on reaping with great accuracy.' When the row of ears of paddy is ripened according to the process due to the consequence of their fate, the irrigation should be ceased – so determined those who best knew the essence of cultivation. This method is asserted by the wise men with regard to all rice-fields. Therefore the cultivators should untiringly protect the eras of paddy from various damages at the time of ripening. In this manner, when the ears at the top of the plants become ripened, their stalks bend their head low to the ground. The cultivator having seen this and having investigated it should protect it himself or get it protected by hired labourers for twenty days. Now, when the stalks of the paddy have become ripe and tinged with golden colour in various places, then the harvest is to be done by the cultivators with sickles (*śaṅkula*) etc. The men should cut the heap of stalks of the rows of paddy in one day or in several days continuously with the help of the dependants and other hired labourers who are all willing to be of assistance. When the stalks are not spoiled or damaged by rain or thieves then the harvest is said to be prosperous.'

In comparison, the KṛP contains a separate chapter on testing the ripeness of paddy before harvesting by cutting off two handfuls and a half and appointing the proper constellations for harvest. What is more, there is a relatively long chapter on the popular ceremony called Puṣyayātrā in the month Pauṣa, i.e. in December-January.

What is common in both texts is that they touch upon such an important procedures as the test for ripening and the protection of the ripe stalks of paddy from damage. The test described in the KṛP (verses 206–213)is quite succint, while that of our text is rather based on a longer observation. Our text does not go into details concerning damage; the KṛP gives a long list of creatures that destroy the crop like insects, pests and beasts like goat, boar, bear, deer, buffalo and birds like sparrow and parrot and offers a higly effective *mantra*, following verse 194, to prevent the lost caused by them. Since the Kṛṣiparāśara keeps entirely silent the actual work of reaping and on the implements involved, it is clear that in this respect the account of our text is more valuable.

Threshing, winnowing

We are given a full account of these procedures in verses 483–490.'The harvesting of different rices is to be carried out in this manner. The cultivators, with the help of hired labourers and attendants, should store the reaped stalks with their ears on threshing floors. Storage and protection on the threshing floor for three to five days is said to be very useful. By threshing the ripe stalks of the heap of rice from which the ears are loosened, the quantity of grain falls on the threshing floor and sparkles like a heap of faultless pearls. And again, when the ears are threshed with effort by the trampling of bull and buffaloes, the remaining quantity of grain also falls on the threshing floor. Having threshed the paddy in this way on the threshing floor the cultivators should pile up the thick, firm and separated grain. Having strenuously separated the non-thick and beautiful by means of a winnowing basket, wind and fanning, the cultivators should pile them up. Then the wise man, having dried them in the sun, should clean them.' This description iindicates that threshing is done in two different ways: with manpower, using a tool not specified here, and by animals and the removing of the chaff from the grain is done with utmost care. Luckily, there is a reference to this tool in verse 563: 'Threshing by a long rod (*daṇḍa*) and the like and the drying in the sun of the harvested ears follows on the threshing floor.'

KA II, 24, 31–33 on the same process is strikingly brief. 'And at the proper time he should bring the grains on (the threshing floor) and other things as they are harvested. The wise man should not leave anything in the field, not even (corn-)stalk. From the heaps he should make high stacks or something like that with tops neither compact nor empty. He should make the stacks at the border of the circle (of the threshing floor). At the threshing floor, his attendants must have not fire but water.'

Shared common features in both texts are the employment of attendants and animals at threshing. The employment of the latter can be inferred from the reference to a circle (*maṇḍala*) in passage 33.

The KṛP does not specify either the tools or the single steps of the procedure of threshing and winnowing.

What is really unique in our text that is the distribution of crops still at the threshing floor in a way known from the so-called *jajmānī* system. Verses 491–492 read thus:' Having given one part as a gift to the gods and the kings according to the rule and one part to the *brāhmaṇa* learned in the *Vedas* according to the rule and one for nourishing the hired labourers, he should store the remaining stock of paddy in his house.'

Similarly, we do not come across the procedure of cleaning pulses in other Sanskrit texts. Verses 568–569 deal with these processes. It's a pity that they contain some obscure terms too. What is clear here it is that the cultivator should collect them on the threshing floor and separate them thoroughly from gravels (*kṣudragrāma*), stones (if we read *grāva* instead of *greva*), and *kaṭu* (?), through the process of throwing them into a winnowing basket, exposing to wind, *pecana* and *cāmana.* Ayachit considers the first as a wrong form of *peṣana* 'pounding' (SE 104). I might associate it with Telugu *pecu* 'clean, peel, rind' (Brown 1980, 971) or would it emend for **pecchana,* a Middle Indic form of Sanskrit *preṅkhana*, which means 'swinging'. *Cāmana* reminds me of the Nepali *cāmal*, an equivalent of Hindi *cāval*, 'husked rice fit for cooking' (Chatterji 1932, 31) and I take the term as 'husking'.

Storing

Verses 493–501 deal with storing grain and straw. 'He should carefully preserve grain in stone vessels, in some places in well-baked clay vessels, in some places in vessels of strong glass or in containers woven of ropes and plastered with mud according to the custom of the region. In some places the cultivator should make a pit in the hard earth, provide it with descending steps and store the grain there and properly protect it from the danger of water, thieves, parrots, rats and harmful animals. The preservation of *kalama, vrīhi, dhānya* and the like and also the preservation of their seeds, which are dried in the sun, is so prescribed by the best cultivators. When the cultivators have taken the heap of paddy to their house, they should gather together the tender straw obtained from the stalks of *sasya, vrīhi* and the like- dried in the sun and trodden over by the bulls – into small piles and store it in their house or in the cow-shed and should specially preserve it.'

After the short treatment of the procedures of reaping and threshing, the KṛP dwells on weighing and especially with the *āḍhaka*, a measuring vessel for grains, in a separate chapter, a topic missing from our text. The last chapter of it deals with storing in a way characteristic of the whole work: it fixes the suitable constellation and provides us with a *mantra*, which kept in the granaries (*dhānyāgareṣu*), ensures great prosperity.

Storing grain is the business of the superintendent of the magazine in the KA II, 15. There is no mention of this procedure in common households.

Animal husbandry

Our text devotes one chapter to the worship of the bull-king, alias *vr̥ṣotsarga* (verses 277–284) and another to the description of the marks of cows and bulls (verses 285–318). It is remarkable that cattle are classified according to the *varṇa system*, i.e. the text distinguishes *brāhmaṇa, kṣatriya, vaiśya* and *śūdra* bulls. The description reminds Nene of the breed known recently by the name Ongole, which is predominant in Guntur and Nellore in Andhra Pradesh (SE 136). Cattle may be kept in cow-sheds or in herds. During the hot season they should be placed in the shadow at noon, but in the cow-pen in the cold season. They must be protected from cow-lifters and from vicious beings. The proper treatment and worship of cattle bestow prosperity upon the cultivators. We learn from verses 501–502 that straw served as fodder for cows, bulls, buffalos, goats, and deers.

The Kr̥P tackles this topic in the chapters on the rule about draught-animals, the description of the festival of cows and the procession and entrance of cows, respectively (verses 84–108). The prescriptions concerning the nourishment are very instructive: draught-animals are to be fed by molasses, fodder (*yavasa*), wheat, and by grazing. The size of the cow-shed has been determined. Cow-sheds must be kept clean, because 'What hope for draught-animals can there be where householders cleanse cow's excrements with cow's urine', or from where these animals go out every day 'with their bodies besmeard with cowdung and cow's urine.'

The KA II, 29 called The Superintendent of Cattle throws light on the subject from the point of the chief official supervising the royal stock of cattle and mainly focusses on the property rights, protection of cattle, dairy productions and the rations alloted to bullocks, mules, cows, donkeys, buffaloes and camels.

Appendix Three

The role of gods in the Kāśyapīyakr̥ṣisūkti

As Nirad Chaudhuri aptly puts it ' religion was a means of living for the Hindus' and 'their religion exercised a rigorous control over the life in every aspect and in every activity'(Chaudhuri 1980, 186). With Gonda's words ' The massses being bound up with external observances and the main form of their religious activity being ritual and ceremonial, the traditional Hindu is convinced that, in view of the fundamental oneness of the universe, and harmonious co-operation of men in ceremonial activities is necessary for the harmonius functioning of the world. All must co-operate if the crops are to be good, and if the life of the village is to be happy' (Gonda 1980, 67–68). It follows from this that a careful analysis of some important events in the life of the agricultural community depicted in the Kāśyapīyakr̥ṣisūkti may be of great interest.

Rites connected with the agricultural year

Digging up a well

Verses 148–149 assume that 'Generally in the countries the lower level [of the ground] the water is low and sometimes even lacking in the river-bed in the summer months. Therefore, digging up of wells and the like in a summer month for the sake of obtaining full supply of water lasting for a long time should be undertaken.' The work of digging proper preceeded by a careful water divination begun in the morning. The first step was the worship of Varuṇa the lord of waters (*nīranātha*) Bhūdevī and Vanadevī with presentation of a *bali* -offering (*balidāna*).

Varuṇa appears here as a god reigning in and over water. In the same way he stands on the list presiding over the eight quarters in the chapter LIV of the Bs. called 'Exploration of water springs' (*Dakārgalam*).

Bhūdevī, Goddess Earth, the personified earth, is the natural source of life and of vitality. She has to be propitiated for being endured harm by digging.

The name Vanadevī, the Goddess Forest, cannot be attested in any other Sanskrit text. Vedic Araṇyānī (RV X, 146), the Goddess Forest, is concerned with with food producing and tending animals, as far as she is 'rich in food without ploughing' and she is the 'mother of animals'. There are goddesses called *vanadevatā* in the works by Kālidāsa: in the Ks. 3, 52 and Rv. 9, 52, but, to my best knowledge, they have nothing to do with water springs or underground water. Crooke reports on a plantain tree, which are associated with fertility, placed to the right of the image of the goddess Durgā, representing Vana Durga [sic!], 'she of the forest', at marriage ceremonies in Sambalpur (Crooke 1926, 411).

The inclusion of the two goddesses to be worshipped and the omission of the ususal gods of the quarters may be due to a peculiar tradition which has come down to us in the KāKS.

As the ingredients of a *bali* offering are not detailed here let us remember that of the Vaikhānasa Viṣṇu rites which are definitely South Indian rites. The ingredients of the later are rice, beans, two sorts of pulse, sesame, symplocos racemosa, cakes, and parched grain and, besides, water, flowers, perfumes, incense, lamps, and unhusked barley-corns. It is conducted in honour of the discus, *śānta* and Garuḍa (Gonda 1993, 251).

Application of a plough and other tools. The commencement of ploughing and other operations

The first task to be done for a man who knows agriculture and who is keen-witted is to meditate on the Trimūrti, the three manifestations of the Supreme being, namely Brahmā, Viṣṇu and Śiva. Harnessing the plough, the cultivator should worship Dhānyadevī with the benedictory sounds of the recital of a prayer. Then having especially worshipped Bhūdevī, Lakṣmī, Vāṇī, Gaurī, Surādhipa, Varuṇa and Dhaneśa, he should start ploughing with bulls and ploughs previously adorned with incense, mango leaves, flowers, sandal paste or saffron.

The name Dhānyadevī, the Goddess Corn, is not recorded in any modern Sanskrit dictionary. She is commonly regarded as a special form of the goddess Lakṣmī; however, here the name seems to refer a goddess whom the compiler of the text, for unknown reasons, wants to distinguish from the goddess Lakṣmī. This usage may reflect a local tradition the details of which await exploration. She has to be adorned with the very auspicious sound of a conch.

Lakṣmī, the goddess of fortune and prosperity, is a tutelary deity of paddy (Gonda 1970, 112). Little wonder that Lakṣmī is represented by a basket or a corn-measure full of paddy on the full moon day in the month *Āśvina* (September – October, the month of harvest). She is also worshipped in a figure shaped from river-sand and and placed in a winnowing-basket (*śūrpa*) on the twelfth day of the light half in the same month (Meyer 1937, 90).

Vāṇī is the goddess of speech, Sarasvatī (MW, p. 939). Sarasvatī is in early times a river-goddess and later of speech and eloquence. She is sometimes considered as the daughter or wife of Brahmā; she is identifed with Durgā or even with the wife of Viṣṇu or Manu (MW, p. 1182). Her association with fertility need not call for explanation. Her particular role here is not clear.

Gaurī lit. 'white', 'yellowish' is the chief Goddess of Harvest (Chattopadhyaya 1968, 233), she is regarded as the source of the world, the Cosmic Cow (Stutley and Stutley 1977, 96). As Payne put it, she was a goddess of the harvest and the ripened corn (Payne 1933, 7). Her feast is celebrated on the 7th, 8th and 9th day of the light half of the month *Bhādrapada* (August-September). This a festival of the early harvest. She is a corn goddess who springs forth the seven sorts of grain to be found in the winnowing-basket, who is incorporated in the seven sorts of grain in the winnowing- basket, the goddess who consists of the fresh seeds shooting up in the winnowing –basket (Meyer 1937, 91). Gaurī may also mean 'red-yellowish' and so denote the complexion of Lakṣmī and refer to the latter, the genius of the ripe corn (Meyer 1937, 93).

Surādhipa, the sovereign of the gods, is Indra who is also the bestower of rain. He stands beside Lakṣmī on the *kojāgara* festival (Meyer 1937, 89) celebrated with various games on the full moon night in month Āśvina (MW, p. 312). Indra as Vāsava is invoked in verse 133 and as Śacīpati in verse 136 of the KṛP at the commencement of ploughing called *halaprasāraṇa*. Indra is among the tutelary gods at the commencement of ploughing in the list of the Brahmapurāṇa quoted in the Gṛhastharatnākara of Caṇḍeśvara Ṭhakkura (GRK, p. 432).

Varuṇa's definite business remains unknown. He does not figure among those gods who are patrons of agricultural activity.

Dhaneśa is the name of Kubera, the god of riches, who is the husband of Śrī or Vasundharā the personified earth (Bhattacharji, 1978, 162), so that he has some connection with agricultural production. It has to be noted that this form of his name is seldom used. His usual names are Dhanapati and Dhanādhipa in the classical Sanskrit literature.Kubera's name is missing from the lists of tutelary gods of agriculture.

All in all, this is a unique list of gods at the application of a plough and at the commencement of ploughing for various reasons. The number of female deities is obviously large in comparison with common texts on the same topic. It may be ascribed to the above mentioned local tradition that may reflect old Dravidian cults where the agricultural deities, for the most part, are female (cf. Whitehead 1921, 17). We are looking for the Maruts, the wind gods, or Prajāpati,the lord of creatures, listed in KṛP 139 and Brahmapurāṇa quoted in the Gṛhastharatnākara of Caṇḍeśvara Ṭhakkura (GRK p.432), or Pṛthu, a cultural hero who taught first the human race to cultivate earth, listed in KṛP 131, in vain.

Plough worship (verses 270–276)

The title 'plough worship' (*lāṅgalapūjā*) is somehow misleading. Quite precisely, it is the description of a special worship paid to Bhūmidevī, the Goddess Earth, before the commencement of ploughing. A free translation of the chapters reads thus.

270. 'Before ploughing, the worship of the plough and especially of the oxen is to be performed by the cultivators who know the favourable signs.

271. The worship of field or land with pure water, incense and the like and moreover with lamps yields fruit.

272. O Goddess Earth (Bhūmidevī), o all-enduring, wide-expanse! Reverence to you! I shall start ploughing. Be pleased o virtuous one!
273. Ploughing, tapping and whatever other harm I did on you, o goddess (Devī) forgive all that and yield me a rich harvest.
274. You are declared to be the mother of all living beings. Therefore be kindly disposed towards me o goddess! Give me a boundless crop on the earth!
275–276. Having thus praised and requested the Goddess Earth (Bhūdevī) – the beloved of Viṣṇu - and making circumambulation with salutation, having praised the guardians of the quarters (*dikdeva*) and the protectors (*rakṣaka*) and the life-promoting sun (*sūryaṁ āyuṣkaram*) he should make the best commencement of ploughing of the land.'

Bhūdevī, Bhūmidevī and Devī are the names of Goddess Earth who is the second consort of Viṣṇu beside Lakṣmī (Gonda 1993, 230. She should be propitiated before ploughing in order that she may endure it and may be benevolent to the cultivators. Besides the lamp is one of the chief accessories of Lakṣmī worship which occupies a central place at the harvest festivals held in the month of *Āśvina*; however its application in the given context seems to be a feature of the tradition reflected in the KāKS. KṛP 139 instructs us thus' Then, for a rich harvest, give without fail an offering and a lamp of ghee to Wind-god (Marutvate) accompanied by cloud.' Goddess Earth is approached with the following incantation in KṛP 140: „O Goddess Earth, with golden womb, resting the serpent called Śeṣa, and supporting the moving and unmoving, grant me the desired food.'

The guardians of the quarters and the protectors are present at almost all ceremonies. Here they are seeked for more security. The term *dikdeva* for the guardians of the quarters instead of the common *dikpāla* or *lokapāla* is noteworthy. The wording *rakṣaka*, 'protector' is not usual. In this sense the words *rakṣa* or *rakṣapāla* can be expected MW, p. 859).

Under the worship of the sun the worship of the Sun-god is meant. It is well-known that the sun is pre-eminently connected with vegetation in various civilizations (cf. Bhattacharji 1978, 227). What is peculiar here is that the role assigned to him here reflects his position occupied in the Vedic times he possess the curative powers and was regarded as the preserver of life. In classical Hinduism he lost this exalted position and became a mythical figure of not so much significance. (cf. Stutley - Stutley 1977, 291). The life-promoting aspect of the Sun-god has been transfered to Viṣṇu. It remains a question what had led the seemingly Vaiṣṇava author/compiler to establish Sūrya in his former position.

The worship of the bull-king

The worship of the bull-king (*vṛṣabharājapūjā*) is an abridged and slightly modified version of an ancient fertility rite has been known as *vṛṣotsarga* the 'setting free' or, rather, 'sending forth of the bull' is a ritualized traditional farmer's custom, the 'wedding' of an excellent bull. The festival has been known first from the *gṛhyastrasūtras* and the ritual manuals belonging to the Atharvaveda. It was held in October-November or April-May, in the months in which the crops are ripe (Gonda 1980, 432–433). For this purpose the finest bull of the herd was chosen. The main events of the ceremony are as follows 'An *ājya* oblation is offered, and a mess of cooked food to Pūṣan by reciting a verse to the deity. When the bull stands among the cows of the herd a verse is recited. With the milk of all cows *pāyasa* is cooked and the brahmins are fed with it' (Sen, 1978, 158–159). A version of the same rite is reported in verses 99–104 of the KṛP under the name *goparva,* the festival of cows where the cultivators take the chief bull round the village with vocal and instrumental music. Our text contains a genuine variant of the rite: the bull-king is regarded as an offspring of Surabhi, the fabulous cow and wife of the mythical sage Kāśyapa; he is worshipped with fragrant garlands of flowers and the like; he may be frutiful by Śaṅkara's (=Śiva) grace; he alone may give the cultivator the growth of wealth and grain. The last verses of this chapter overlap the relevant parts of the KṛP that deal with the proper treatment of the cattle. While the description of *goparva* and the rules of keeping of cattle preceed the passages depicting the act of preparing the plough and its accessories for ploughing in the KṛP, *vṛṣabharājapūjā* is placed after those processes. The obvious reason for that may be the loose edition of the whole KāKS. Less probably this may be due to a special tradition preserved in our text.

Worshipping gods at the storing of paddy

In chapter 13 we are given a lengthy and subtle description of the whole process of producing paddy at the end of which there are ritual hints with regard to the storing it. It reads so: 'By the advive of the nobles (*ārya*), by the custom of the country *brāhmaṇas, kṣatriyam vaiśyas, śūdras* and men of mixed origin who are devoted to agriculture and others who possess land after having stored the mass of ripe paddy in their own houses at a lucky hour and at an auspicious moment and after that having worshipped Dhānyadevī and Dhanalakṣmī with incenses, flowers and pure substance – determined by the textbooks – and having fed the *brāhmaṇas* and paid them the proper fee, they gain glory, divine favour in this world and will always happily reign (*rājanti*) united with their good sons, wifes and grandsons.' (verses 505–509)

Dhānyadevī, the Goddess Corn, and Dhanalakṣmī are the wifes of Viṣṇu. These names cannot be attested in other Sanskrit texts. The passage is of high importance, because it reflects a tradition where fine distinction has been made between Lakṣmī and Gaurī, Lakṣmī and Bhūdevī etc. and Lakṣmī and Dhānyadevī while the *mantra* and verse 240 of the KṛP prescribe the worship of Lakṣmī, the giver of wealth only.The name Dhanalakṣmī may seem to a pleonasm too; however this rather artificial form may also be used on purpose to express unambigously the two distinct aspects of Viṣṇu's wife.

Rites connected with the agricultural year missing here

Āgrayaṇa

A rite called *āgrayaṇa* i.e. 'oblation of first-fruit ' to gods has been known since the age ofthe Śatapathabrāhmaṇa (Mylius 1974, 433–442) It is frequently mentioned in the various Śrautasūtras (Sen 1978, 75). The MaS VI, 10 records it among the great sacrificies. The BhāP XII, 21, 48 depicts the festival with few touches: 'Enriched with ripe corns the earth looked happy with (Vedic) sacrificies carried on with oblations of first fruits at the end of the rainy season as well as with (other) grand festivals conducted for the gratification of the senses in cities and villages, and more so with Śrī Kr̥ṣṇa and Balarāma (the two Avatāras of Śrī Hari).' (C.L. Goswami's and M.A. Śāstri's translation) Verses 206–213 of the Kr̥P treats it under the name *agrahāyana* 'taking a handful (of paddy)' which secures good harvest and wealth.

Sītāyajña

This is a sacrifice to the goddess Sītā, the personified furrow. It is performed on a ploughed field to secure protection of the crop where a mess of food is cooked with barley or rice grains, and offered to Sītā, Yajā (the goddess of sacrifice) and Bhūti (the goddess of devotion)in the Gr̥hyasūtras (Sen 1978, 1659). According to the KA II, 24, 27 at the first sowing the cultivator should recite the following *mantra*: 'Salutation to Prajāpati and the god (?) Kāśyapa always. May goddess Sītā prosper in my seeds and goods!' All this is missing from the passages dealing with the collection of seeds and sowing in our text. It is rather strange because god Kāśyapa, a genius of agriculture, appears on the list of the Brahmapurāṇa quoted in the Gr̥hastharatnākara of Caṇḍeśvara Ṭhakkura (GRK, p.432) A certain Udālaka Kāśyapa is invoked at the time of harnessing a plough in the PāGS II, 13. Kāśyapa was the genius who made the Kashmir valley fertile (NīP 300 and Rt. V, 109 and 113). Instead of it there is a dialogue in the style of the Purāṇas between the Goddess Earth variously called Bhūdevī, Bhūmidevī, Dhātrī, Devī Medinī and the Creator under the names Brahmā, Dhātr̥, Kamalāsana (that who has a lotus as seat), Lokapitāmaha (the paternal grandfather of the world) and Vidhi where the former simply states that there is no doubt that all the various seeds are created by Brahmā and carefully sown in her necessarily have sprouts very soon. It remains a puzle for the time being why our text bearing the name of Kāśyapa in its title has not kept this line of tradition.

Rites concerning cooking and food

Worship of gods before cooking

The *brāhmaṇas, kṣatriyas* and *vaiśyas* should bring in their houses pure substances like rice, vegetables, water, milk, curd, ghee, butter and oil which are clean and fresh. The they themselves or theit wives must clean the pots, smear the cooking floor with cow-dung and cleanse the fireplace endowed with fuel. After that they should worship the earth, Vāsava, Yama, Agni, Varuṇa and Brahmā.

Vāsava (Indra), Yama, Agni and Varuṇa figure here likely as the guardians of the quarters. Because descriptions of cooking are very few in Sanskrit literature and even these accountss do not treat this ritual part of cooking, this account seems to be unparalleled.

Worship of gods before eating

Verses 808–809 declare the following: '*Brāhmaṇas, kṣatriyas, vaiśyas, śūdras* and others also, having known the order of eatable and unetable things according to the custom of the country eat the food which is offered first to god and then they become sinless and adorned with the virtue of faith and devotion.' The seventeenth century Bhojanakutūhala, 'Curiosity for food Bhok., p.215 lines 15–21) prescribes that Brahmā, Viṣṇu, Maheśvara (Śiva) and Hanumān should be remembered at the commencement of eating. The seventeenth-eighteeenth century ŚRK VI, 20, 42 states that one having delivered the food to gods on a dish may take it for himself. The food is considered as part of Brahmā (cf. Stevenson 1971, 242) therefore he should especially be worhipped.

Food offerings to gods

Section four is devoted to the preparation of a great variety of foods to be offered to Viṣṇu in his various forms and *avatāras*. The text distinguishes three kinds of food namely *bhojya*, that is fit for guests and co-citizens, *kavya* meant for the manes and *havya* (oblation) to be offered to gods. It is also stated that gods should be worshipped in their images: Viṣṇu in the form of *śālagrāma*, Śiva in the shape of the *śivaliṅga*. The offering to be placed in front of gods should be the best food of various ingredients, embellished by various condiments and cooked by righteous ladies. The employment of the latter is recommended in section three (verse 782) dealing with the rules of cooking ritually pure food for *brāhmaṇas*. This custom is unknown in the *smṛti* literature, but it is common in Tamil nadu even in our days.

The list of foods to be offered to Viṣṇu is relatively rich. Some items of the list are popular overall in India such as *apūpa* 'round cake made of pulse fried in oil or butter', *pāyasa* 'rice boiled in milk' and *vaṭaka* 'balls prepared with black gram soaked in water and ground on a piece of stone and fried in oil or ghee. Rice and *temana,* a soup made from curd (SE 144), ann various foods cooked with faith and devotion and food cooked under certain conditions from pure substance in the kitchen are also fit to be offered.

There are rice dishes that do not occur in any other Sanskrit text are *guḍānna* 'rice mixed with dry sugar', *tilānna* 'rice mixed with sesame', *dadhyānna* 'curd-rice' and *mudgānna* 'rice with *mudga* bean (Phaseolus radiatus Linn.). There are some special sweetmeats like the for the time being unidentified *ḍala* and *mahāḍala and mājaka* unattested elsewhere. Besides the text contains the other *hapax legomena*: *godhūmavaṭaka ' balls prepared with whet and fried in oil or ghee, payogalabandi/payogulabandi* round sweetmeat prepared with milk and dry sugar, *śarkarābandi* 'roud sweetmeat prepared with milk and dry sugar'.

The *brāhmaṇa* should place all this mass of food according to the rule in front of Viṣṇu. He should accomplish the worship with flowers, incenses and ointments, fanning chowries and chanting hymns of praise. He should perform all this worship like the royal service both in daily and occasional worship diligently. He should praise Viṣṇu in his various forms such as *Jagannātha* the lord of the world Kṛṣṇa and Paramātmā, the Supreme Soul who appears here to be Viṣṇu- Kṛṣṇa.

This representation of food offerings to gods has many unique features. As a whole, it presents a highly theorethized normative system which ought to be valid for the religious practicioners of a *Vaiṣṇava* rural community.

Appendix Four

Edible and inedible in the Kāśyapīyakṛṣisūkti

Although there are references to the science of cookery (*sūpaśāstra*), no ancient texbook on cookery has survived (Basham 1981, 215–216). R. P. Das in his critical review of an edition of the Pākadarpaṇa remarks that the content of this book differs amazingly from those of books on South Asian cookery (Das 2001, 122). The extant texts under the headings *pākaśāstra* and *sūpaśāstra* apparently fall in the category of *āyurveda* i.e. the science of medicine. Cooking, or making tasty food, was originally an art willingly undertaken even by kings, In other words the science of cooking was originally a part of the *arthaśāstra* studies and cultivated even by kings. As Meulenbeld puts it 'many monographs were written on this subject (*pākaśāstra*), but they have hardly been studied. The number of *pākas* in therapeutic treatises increases in later periods of Indian medicine (after the sixteenth century?)' (Meulenbeld 1984 66, n. 70). Practical cooking, of course, existed in parallel, however, not under the above mentioned names.

There is a sheer paucity of data on this subject in the *arthaśāstra, dharmaśāstra* texts and the classical Sanskrit literature. The term *piśitapākopadeśa*, 'instruction on how to cook meat' and the qualification of a cook (*pācaka*) 'who has the true knowledge of the modes of mixing to embellish with all favours' *(sakalarasapasśādhanavidhivyatikarādhikavivekaḥ*) in the Yt. II, 126, 7 are allusive.

In light of all this, the data on culinary arts preserved in our text are of great value. These data are of two kinds: incidental passages inserted in various chapters and coherent passages in section three called 'The Description of Rules about Edible and Inedible Things' (*bhojyābhojyakramakathanam*) and section four called 'The Description of the Order of Making Various Oblations'.

Edible things

To begin with incidental passages, there is a brief excursus on healthy nourishment in chapter 14 dealing with growing pulses. This digression contains some important references to the ancient tradition of culinary arts.

Verses 521–526 read thus: 'The seeds of the pulses *āḍhaka* (Cajanus india Spreng.), *māṣa* (Vigna mungo (L.) Hepper), *caṇaka* (chick pea,Cicer arietinum L.) *mudga* (mung bean, Vigna radiata (L.) Wilczek), *kulattha* (horse gram, Dolichus uniflorus Lam.), and *tīla* (Sesamum indicum L.) are to be sown and *marīci* (black pepper, Piper nigrum L.) is to be cultivated. Sauce is said to be made from them. In this world, various foods are made from rice, which comes from the paddy for the sake of protecting life. The *pākaśāstraka* (the textbook on cooking or the science of cooking) gives instruction concerning this matter. King Vīrasena composed a big *pākaśāstra* and his son Nala and the great sage Garga did likewise. Various series of foods and sauces are mentioned there. Those sauces make the most delicious food which are made from *māṣa, āḍhaka, marīca* and salt, and these prove to be nourishing and health-giving to human beings.'

Vīrasena appears to be a misnomer for Bhīmasena i.e. Bhīma who disguised himself as a cook in MBh IV, 2, 1–7 and who, according to the BKS XVI, 61, was an expert in *sūdaśāstra*, the science of cooking'. Vīrasena has not traditionally been known as an excellent cook, although he was indeed the father of King Nala. Nala authored the book called Pākadarpaṇa 'The mirror of cooking', a relatively late text on Indian dietetics (Das 2001, 121). According to the MBh III, 72, 12, Nala was well versed in *sūdatva,* 'the business of a cook (MW 1242). The Rasadīpikā commentary on Km. 124 explains the term *sūdaśāstra* as *nalādipākaśāstra*, 'the cooking book composed by authorities beginning with Nala'.

Garga is the name of an astronomer or of a physician (MW 349). Daśarathaśāstrī, son of Nārāyaṇa and the author of the Kṛṣiśāsana, a twentieth century Sanskrit treatise on agriculture calls himself a descendant of the Garga clan (Gargagotra) of *brāhmaṇas.* The relation of the above mentioned persons to cooking cannot be asserted by the available textual evidence.

Verses 611–614 say: 'The first to derive from the agricultural undertaking is the rice ingredient, the second is the sauce ingredient, the third is the vegetable ingredient. Ghee, milk and curds are mentioned as the fourth ingredient. These four ingredients make the food-stuff. This latter increases the pleasure of Gods and provides life stability for all men here on Earth. This gives nourishment and health, increases intelligence and age of life. This was created by Brahmā throughout all lands at the time of creation of the world.'

Verses 811–815 instruct us that proper food should be made from pure substances, rice and the like and vegetable substances which are cleaned and from milk, coagulated milk, ghee, fresh butter and oil. Food prepared according to certain rules should be adorned with the six *rasas.* The six *rasas*, different tastes, are as follows: sweet (*madhura*), sour (*amla*), salty (*lavaṇa*), pungent (*kaṭuka*), bitter (*tikta*) and astringent (*kaṣāya*) (MW 869).

Verses 837–838 say:'Rice-food, which is cooked in oil or ghee for the proper time, and the sauce is very tasty, as well as the various foods prepared with faith and devotion and food cooked under certain conditions from pure substances in the kitchen.'

All these prescriptions reflect strict vegetarianism. But this is compulsory only for the *brāhmaṇas.* Verses 806b-807 make it clear that 'The sages who see the truth have not declared this restriction for *kṣatriyas, vaiśyas* nor *śūdras* in this world. Even the eating of meat is not sin for them – it is thus determined.' This is a general statement while similar ideas of the KṛP 222–224, which allow both vegetarian food as well as fish and meat at the *puṣyayātrā* festival, following harvest, must be due to the local customs in Bengal.

Inedible things

Verses 782b – 806a of this Section, pertaining to forbidden food (*abhakṣya*) and unfit food (*abhojya*) (Olivelle 2002, 13–14), precisely regulate the method of cooking for the *brāhmaṇas* or by the *brāhmaṇas.* 'Therefore the pious *brāhmaṇa,* knowing the rules about edible and inedible things among manifold ingredients, should employ a righteous lady for cooking. Food prepared from *pilu* (mustard tree, Salvadora persica L.) and *kodrava* (*kodo* millet, Paspalum scrobiculatum L.) is unfit for Gods and *brāhmaṇas.* As a rule, *śubha-* and *śvetavārtāka* (eggplant, *briñjal*, Solanum melongena L.), *laśuna* (garlic, Allium sativum L.), *gṛñjana* (shallot, Allium ascalonium L.), *palāṇḍu* (onion, Allium cepa L.), *kavaka* (mushroom, Agaricus campestris L.) and also the exudation of trees are not to be eaten by *brāhmaṇa,* nor are products from dirty places. The *brāhmaṇa* should avoid *bṛhatkūṣmāṇḍaka* (big pumpkin - gourd, Benincasa hispida (Thunb.) Cogn.or B. cerifera Savi), *pannā* (?), *śigru* (radish, Raphanus sativus L.), *piluka* (mustard tree, Salvadora persica L. or toothbrush tree, Careya arborea L.) and products from improper places.'

These plants are regarded as *jātiduṣṭa* items i.e. forbidden on account of their very nature (Kane 1974, 771). Verse 786a points to a *kriyāduṣṭa* prohibition i.e. forbidden on account of certain actions improperly done (Kane 1974, 771).

Pilu or *piluka* is a shrub or small tree with a short twisted trunk, and drooping branches. It is generally considered as a medicinal plant, but it is also used as a potherb (P.V.Sharma 1979, 62–63). Its berries are considered hot and digestive. The obvious reason of prohibition is not clear. *Kodrava* is a kind of paddy 'said to be powerfully narcotic and is eaten only by the poor who prepare it in various ways' (Meulenbeld, 1974, 546). *Śubha-* and *śvetavārtāka* are the *śubha* (lit. 'shining, beautiful') and the dark varieties of eggplant. It does not appear on the list of *abhakṣya* vegetables in the well-known *smṛti* texts. According to Sen Gupta it counts a prohibited item in the Uśanassmṛti and the Paithinasīsmṛti (Sen Gupta, 1956, 190). *Laśuna, gṛñjana, palāṇḍu, kavaka,* and the exudation of trees figure in MaS V, 5, 7, VāDhS 14, 33, VaiSmS IX, 15 and the PaP 6, 253, 110–112.

To the best of my knowledge, *bṛhatkūṣmāṇḍaka* is not recorded as a prohibited item elsewhere. *Pannā* is an unidentified name, or more likely a corrupt reading in the text. *Śigru* is normally edible. Its consumption is strictly prohibited only by YāS I, 171 and 176 for the *brāhmaṇas.*

Verse 781a states that 'From the teaching of gods cooking should be made by pure women.' Likewise, verses 812b-813a declare that both the cleaning of the cooking stove with cow-dung and cooking should be made by the *brāhmaṇas* or by their own pure wives (*svīyapatnibhiḥ sādhvībhiḥ).* The employment of women as cooks is common in South India nowadays. This usage cannot be traced in

Sanskrit sources. The word *pācakabhāryā*, 'having a cook for a wife' P VI, 3, 37 Sch. quoted after MW 614, conveys to us simply the idea that the wife cooked under certain circustances. V. S. Agrawala mentions, in connection with the not extant Sūdaśāstra by Bhīmasena, that it was studied by the female cooks attached to royal kitchens (Agrawala 1974, 321). Unfortunately, he does not make reference to any textual evidence, therefore it seems to be only a conjectural hypothesis. At least, the story of Gominī in chapter six of the Daśakumāracarita is about a girl who quickly prepares a delicious dish from rice in the presence of Śaktikumāra, a young merchant who puts her cooking skills to test (DKC pp. 161–162).

Verses 786b-796 read thus: 'The twice – born should strenuously avoid the food which was prepared, in the house, in a pot defiled by ants, flies and worms. The *brāhmaṇa* should avoid food seen by dogs and cocks etc. and also which is given untrustworthily and defiled by nails or by chaff or fibers of husk. The *brāhmaṇa* should strenuously avoid that food which is defiled by cats, smelt at in the house by muskrats and mice. One who has a superior aim should not, in general, eat food which has been defiled by a low man or smelt at or touched by such a person. And, he should avoid all food which is given in an impure place. For the sake of glory and fulfilment of welfare, the sages said that food remainings from a mouthful, or fallen from the mouth was not to be eaten. Similarly the food that was not [first] offered to guests or Gods because it was given at an improper time, and food that has been prepared in the house in a dirty place, or food not washed by water should not been eaten. Edible (*bhakṣya)* food and vegetables for chewing (*khādya*) are determined inedible for *brāhmaṇas* if prepared by a man who is not of a good family, or who has been defiled by dirt and urine. The *brāhmaṇa* should avoid food which has been defiled by hair, from which he himself had formerly eaten [a bit and put aside or according to Ayachit: that which is eaten first by the cook himself (SE 117)], which has been prepared by an unwashed man.'

These prohibitions belong to the category *saṁsargaduṣṭa* i.e. foods spoilt by foul contact (Kane 1974, 771), *kriyāduṣṭa, jātiduṣṭa* and *parigrahaduṣṭa* i.e. 'food which may be good in itself, but is forbidden because of its coming from the hands of certain persons (Kane 1974, 786).

MaS IV, 207, BauDhS II, 7, 7 prohibit food in which an insect is found; VaiSmS IX, 15 declares food spoiled by worms and insects. The defilement by a dog and cock is mentioned in the MaS III, 239 and by a dog in the VaiSmS IX, 15; MaS XI, 160, VāDhS XXIII, 11 and ViS LI 46–47 call somebody who eats what is left by a cat or mouse punishable. The defilement by a low man (*nīca*) can be compared with the same by a *caṇḍāla* in the MaS III, 239, GDhS XV. 24, ĀDhS II, 17, 20 and ViS LXXXI, 6–9. VaiSmS IX, 15 says that an unwashed man spoils food.

Verses 791–801a say: 'The best of twice – borns should avoid food given by hand, given with the sacred thread hanging down towards the left part and over the right shoulder (*apasavyakaradatta),* given in a broken iron vessel, or a broken wooden box, stale, which has been prepared in another house and then brought to his house. That is determined inedible by the sages who have studied the textbooks. The twice – born should not eat boiled rice with sour coagulated milk and rice with sesame in the night. Eating by moonlight or in a place which is without lamp or eating in darkness is to be condemned – this is determined by the best sages and also Gods.'

The items listed in verse 797 can be labelled as *kriyāduṣṭa* prohibition. Verses 798–801a speak of *kāladuṣṭa* prohibitions i.e. foods forbidden because of the inappropriateness of the time (Kane 1974, 771).

The same spoiling effect of a broken vessel is recorded in the VaiSmS IX, 14. Stale food, as such, should not be eaten (Kane 1974, 784). Food prepared in another house and then brought to the house of a *brāhmaṇa* is prohibited in the VaiSmS IX, 14. Eating boiled rice with coagulated milk and rice with sesame in the night is prohibited in the VaiSmS IX, 15.

Verses 801b-802:'With effort, the wise man should avoid foamy water, water directly taken from a river etc., water fallen from the sky, the water of a current or which has been touched by an untouchable.'

These prohibitions conform to the *jātiduṣṭa* category. I could not find such a systematic regulation of drinking water elsewhere. Yamasmṛti, quoted in the GRK p. 338, forbids the consumption of water from the outlets of wells and the like, however, permits the drinking of water from a river.

Verses 803–806a: Camel's milk, milk of whole-hoofed, or cloven-hoofed animals, salty milk, the milk of a young woman, the milk of a cow without calf, which has been issued after a calf, goat's milk, milk poured in copper: these are known as censurable and are to be avoided. Therefore, the learned pious *brāhmaṇa,* having known the rules about edible and inedible things, should do the cooking according to the rules for the sake of success in sacred study.'

The first four items and eighth belong to the *jātiduṣṭa* category; the fifth to seven ones fall in the *kāladuṣṭa* category; the last one to the *saṁsargaduṣṭa* category.

Camel's milk and the milk of whole-roofed animals are prohibited. The same rule holds true in the MaS V, 8, GauDhS XVII, 32–33, ĀpDhS I,17, 22–24, BauDhS I, 12, 9–11, VāDhS XIV, 34–35, ViS LI, 38–40, YāS I, 170 and VaiSmS IX, 15. The prohibition concerning clove-hoofed animals and salty milk are unparalelled. The milk of a young woman is similarly prohibited in the MaS V, 9. The milk of a cow without a calf is prohibited in the MaS V, 8, that of which has been isssued after a calf (ib.) and in the VaiSS IX, 15. For goat's milk, I did not find any rule in *smṛti* texts. *Smṛti* texts either permit or prohibit the use of copper vessels in households. Kane says – without reference to his source – that instead of them, vessels made of bell-metal (*kāṁsya*) may be used (Kane 1974, 761). This latter rule is rather questionable: verse 90 of the KṛP clearly states that the keeping a bell-metal vessel in a cow-pen is harmful to the animals.

Verse 805b-806a: 'Therefore the *brāhmaṇa* who is learned in the Veda and knows the sacred law about edible and inedible things should do the cooking according to the rule for the sake of success in sacred study.'

The governing verses in our text are *varjayet* 'he should avoid' and its prefixed forms, such as *vivarjayet* or *parivarjayet.* The same verb is used in the YāS I, 170. Other forms indicating prohibition are *abhakṣya, abhojya, varjya* and *nindya.* Similarly to MaS IV, 207, the optative *bhuñjīta* with the negative particle *na* serves the purpose of prohibiting the consumption of special foods. All these forms are connected with *abhojya* food.

The list of plants figuring as food ingredients, which are strictly prohibited, are called here *abhakṣya,* like in *smr̥ti* texts.

Our text puts some prohibitions into the *abhakṣya* category that may fall in the *abhojya* category as it is formulated by Olivelle: the prohibition of food remaining from a mouthful, or fallen from the mouth and food prepared by a man who is not of a good family, or food which has been defiled by dirt and urine. These latter ones have to do with the pure/impure distinction. The question is whether this usage of the term *abhakṣya* is deliberate or not. The deliberate usage would mean that our text differs from the general standpoint and represents an independent tradition. The second possibility cannot be ruled out. It can be ascribed to a kind of poetical license of the compiler of our text.

The prohibitory rules are generally in accordance with the instructions of the respective *smr̥ti* literature. In addition, there are some pecularities among the prohibitions: items such as *br̥hatkūṣmāṇḍaka* and *pilu(ka)*; the defilement by nails, chaff, fibers of husk; food remaining from a mouthful or fallen from the mouth; by giving with the sacred thread hanging down towards the left part and over the right shoulder; use of broken iron vessel or broken wooden box; the detailed prohibition with regard to drinking water; the milk of cloven-hoofed animals and that of goats. Other prohibitions figuring here are obviously quite practical, though not fixed in such details in *smr̥ti* texts: given in an impure place; given at an improper time; prepared in the house in a dirty place; food not washed with water; prepared by a man who is not of a good family; defiled by dirt and urine.

It appears that our text heavily draws on the MaS and the VaiSmS. The intricate relationship between MaS and the *Vaikhānasa* texts has not yet been clarified. Caland believes that MaS did probably know the VaiSmS (Caland, 1927, XVIII). The MaS can safely be dated back to sometime between the second century B.C. and A.D. second century (Olivelle 2004, XXIII). At the same time, the absolute chronology of the VaiSmS is not fixed. Therefore I think that the KāKS independently borrowed from the both. A fine example is the instruction that the twice – born should not eat boiled rice with coagulated milk and rice with sesame in the night. Nene finds it unexplained (SE 143). It cannot be found in any text of the *smṛti* literature. However, there is an almost similar passage in the VaiSmS IX, 15. It reads: 'sesame, barley - meal, sour coagulated milk and roasted grain must not be eaten at night' (Caland 1927, 213)

The great number of borrowings from the VaiSmS strongly advocates a possible South-Indian provenence of the extant text of the KāKS.

The employment of female cooks, which can be a modern interpolation to the text, seems to point to the same provenance.

These borrowings, together with the above peculiarities, form a rich stock of material on food prohibitions which is notably distinct from the picture drawn by the well-known *smṛti* texts. This is a fact that makes the passages under discussion particularly interesting, moreover, important.

A salient feature of our text is the exuberance of passages concerning food prohibitions. As to the proportion of the length of treatment of edibles and inedibles, our text is in accordance with the respective *smṛi* and *Vaikhānasa* texts. Nevertheless, it surpasses them in certain details and statements which may be based on local tradition.

Appendix Five

A list of Sanskrit names of plants in the Kāśyapīyakr̥ṣisūkti with their English and botanical equivalents and place of occurence in the text

agastya Sesbania grandiflora (L.) Pers. 578.
atimukta Ougenia oojeinensis (Roxb.) Hochr. 390. = *tiniśa*
arjuna arjun, Terminalia arjuna (Roxb. ex DC) Wight et Arn. 396, 742.
aśoka asoka tree, Saraca asoca (Roxb.) De Wilde 393, 746.
aśvatthaka the holy fig tree, Ficus religiosa L. 91. = *caladala*
āḍhakī pigeonpea, Cajanus cajan (L.) Millsp. 570, 701, 565, 566 = *āḍhaka* 329, 336, 338, 521, 528, 529, 532, 554, 572, 594 and 603. A variety called *br̥hadāḍhaka* 'big pigeonpea' 538.
āmalaka emblic myrobalan, Indian gooseberry, Emblica officinalis Gaertn. 392 = *āmalakī* 744.
āmra mango, Mangifera indica L. 102, 633, 665 = *rasāla* and *rasālaka*
ārdra ginger, Zingiber officinale Roscoe 671 = *ārdraka* 646 = *śuṇṭhī.*
āvali coriander, Coriandrum sativum L. 103.
ikṣu sugarcane, Saccharum officinarum L. 381, 584, 586, 592, 631, 671 = *puṇḍra*
udumbara country fig tree, Ficus racemosa Wall. 395.
eraṇḍa the castor-oil plant, Ricinus communis L. 383, 742 = *eraṇḍaka* 540.
elā cardamom, Elettaria cardamomum Matton 389, 608, 630, 651, 665, 671.
kaṭa clearing nut tree, Strychnos potatorum L.f. 386.
kadamba kadam, Anthocephalus chinensis (Lam.) A. Rich ex Walp. 92, 742 = *nīpa*
kadalī banana, Musa paradisiaca L. 388, 422, 581, 589, 592, 655, 665, 631, 671 = *mocā, rambhā.*
kanda Amorphophallus campanulatus (Roxb.) Bl. Ex Decne. 387 = *sūraṇa*
kapittha wood -apple, Limonia acidissima L. 394, 746.
karañja Indian beech, Pongamia pinnata (L.) Pierre 391 = *karañjaka* 742 = *naktamāla*
karṇikā Indian white rose, Rosa alba L. 747.
karṇikāra Pterospermum acerifolium Willd. 745.
kalama rice sown in May and June, Oryza sativa L. 356, 358, 362, 363,425, 444, 454, 458, 497.
kalāṭa Problematic. It can probably probably be identified with Ceylon gamboge (?), Garcinia morella (Gaertn.) Desr. (?) 386, 646 = *kalāṭaka* 605. It may be

a wrong form standing for *kalāya* pea, Lathyrus sativus L. the leaves of which are used as potherb (Sharma 1979, 104).

kavaka mushroom, Agaricus campestris L. 784.

kārpāsavṛkṣa cotton tree, Gossypium arboreum L. 599 = *tūlāvṛkṣa, picuvṛkṣa.*

kāśa thatch grass, Saccharum spontaneum L. 397, 733.

kīraka ? 380.

kuṭaja a smalltree with white flowers (Randhawa 1980, 507), Wrightia tomentosa (Roxb.) Roem. et Schult. 639–

kunda jasmine, Jasminum multiflorum or pubescens Willd. 103, 390, 639.

kuravaka red amaranth, Barleria prionitis L. 747.

kulattha Madras horse-gram (Das 1988, 306), Dolichus biflorus L. 521. = *kuluttha* 59 = *kulastha* 381.

*kuśa*grass with long pointed stalks, Desmostachya bipinnata (L.) Stapf. 733.

kustumburu coriander, Coriandrum sativum L. 605 = *dhānyaka.*

kūṣmāṇḍa pumpkin-gourd, Beninkasa hispida (Thunb.) Cogn. 386, 605, 630, 636, 646, 651, 659, 666. It varieties: *bṛhatkūṣmāṇḍa* big pumpkin-gourd 785 and *vanakūṣmāṇḍa* wild pumpkin-gourd 630.

kṛṣṇasāri Problematic. It stands on the list of pulses. 381.

kodrava kodo-millet, Paspalum scrobiculatum L. 380, 539, 576, 577.

kolakandāvali It is probably identical with *kolakanda* Urginea indica Kunth. (Meulenbeld) 91.

kovidāradruma St. Thomas tree, Bauhinia purpurea L. 749.

kramuka areca-nut, Areca catechu L. 389.

khadira catechu, Acacia catechu (L.f.) Willd. 91.

khādira Problematic. It may be identical with *viṭkhadira* Acacia farnesiana Willd. or *arimeda* Acaccia leucophloea Willd. (Meulenbeld) 91.

gṛñjana garlic Allium sativum L. 784 (*grañjana*)

godhūma emmer (wheat) Triticum dicoccum Schübl. 380, 539, 554, 566, 570.

canaka chickpea, Cicer arietinum L. 336, 380, 521, 529, 539, 554, 556, 594, 601, 701.

candana sandal, Santalum album L. 730 = *varṇaja*

*campaka*golden champa Michelia champaca L. 103.

caladala the holy fig tree, Ficus religiosa L. 395. = *aśvatthaka*

cāmpeya iron-wood, Mesua ferrea L. 390, 639 = *cāmpeyaka* 747.

jaṭikā nutmeg(?), Myristica fragrans Houtt. 384, 604, 620. The copyist of the ms. curiously identifies it with Tamil *avarai* filed-bean, Dolichus lablab L. (Vacek), however, this interpretation cannot be proved on the basis of any Sanskrit text or dictionary.

jaṭilā long pepper, Pipper longum L. 637. It may denote other plants too (MW 409).

jambīra citron tree, Citrus medica L. 396, 651, 732.

jambu rose apple, Eugenia jambolana lam. 394, 633, 746.

jīra cumin, Syzygium cumini (L.) Skeels. 380, 539, 570 = *jīraka* 382, 540.

tamāla Garcinia morella Desr. (Meulenbeld). It can be identified with other plants (Das 1988, 168–169) 393, 747.
tāpiñcha kokam butter tree, Garcinia indica (Dup.Tho.) Choisy. 747.
tāmbulavallikā betel, Piper bettle L. 671 = *nāgavallī, pūga, pūgikā.*
tiniśa Ougenia oojeinensis (Roxb.) Hochr. 250, 395. = *atimukta.*
tintriṇī tamarind, Tamarindus indica L. 92, 748.
tinduka Diospyros peregrina (Gaertn.) Gürke 92, 250, 396, 743.
tila sesame, Sesamum indicum L. 380, 521, 529, 531, 539, 559, 570 = *tilaka* 396.
tulasī sacred basil Ocimum tenniflorum L. 733.
tūlā Asiatic cotton, Gossypium herbaceum L. 383, 540.
tūlāvr̥kṣa cotton tree, Gossypium arboreum L. 395 = *kārpāsavr̥kṣa, picuvr̥kṣa.*
tr̥ṇa grass 446, 447, 448.
dāḍima pomegranate, Punica granatum L. 393, 746.
devadāru Indian cedar tree, Cedrus deodora (Roxb. ex Lamb.) G. Don. 391, 743.
drākṣā vine, Vitis vinifera L. 389, 608, 651, 652, 665, 702.
dhānya rice, Oryza sativa L. 382, 497, 498, 527, 572 = *kalama, vrīhi,śambaka, śāli.*
dhānyaka coriander, Coriandrum sativum L. 386, 636 = *kustumburu.*
naktamāla Indian beech, Pongamia pinnata Pierre. 391 = *karañja, karañjaka.*
nalada cinnamon, Cinnamomum zeylanicum Blume. 608, 630.
nāgavallī betel, Piper bettle L. 631 = *támbūlavallikā, pūga, pūgikā.*
nārikela coconut tree, Cocos nucifera L. 633, 660, 665.
nimba neem, Azadirachta indica A. Juss. 92, 392, 745.
nīpa yellow teak, Adina cordifolia (Roxb.) Hook f. ex Brand. 396, 742.
nīvāra wild rice, Hygroryza aristata Nees. 366.
paṭolikā small cucumber, Trichosanthes dioica Roxb. 384, 604, 626, 636, 645, 651, 657 = *paṭolī* 663.
panasa jackfruit, Artocarpus heterophyllus Lam, 102, 632, 659, 665.
palāṇḍu onion, Allium cepa L. 608, 784.
palāśa bastard teak, Butea monosperma (Lam.) Taub. 395.
pāribhadra coral tree, Erythrina variegata L. 749 = *pāribhadraka* 395.
picuvr̥kṣa cotton tree, Gossypium arboreum L. 595 = *kārpāsavr̥kṣa, tulāvr̥kṣa.*
piṇḍītaka Randia uliginosa DC. 392.
pippalī the long pepper plant, Piper longum L. 672, 732.
pīluvr̥kṣa mustard tree, Salvadoria persica L. 396, 749.
puṇḍra sugarcane, Saccharum officinarum L. 381, 581 = *ikṣu.*
punnāga Indian laurel, Calophyllum inophyllum L. 396, 665.
pūga betel, Piper bettle L. 665, 671 = *pūgikā* 631 = *tāmbūlavallikā, nágavalli.*
prasavaka Buchanania latifolia Roxb. 98.
priyaṅgu millet, Calicarpa macrophylla Vahl. 392, 744.
plakṣa Ficus lacor Buch.-Ham. 395, 743.

badara jujube, Zizyphus mauritiana Lam. 396.
bilvaka Bengal quince, Aegle marmelos Correa ex Roxb. 733.
bhadraka long-leaved pine, Pinus Roxburghii Sarg 746 = *bhadradāru*
bhadradāru long-leaved pine, Pinus Roxburghii, Sarg 744 = *bhadraka*
bhūrjaka Indian birch, Betula alnoides Buch. Ham. 92 (wrongly: *bhūrjara*), 391, 742.
madhuka mahua, Madhuca indica J. F. Gmel. or mowra fat, Madhuca longifolia (L.) Macb. 395 = *madhūka* 749.
marīci pepper shrub, Piper nigrum L. 382, 540, 572, 620 = *marīca* 522.
maruvaka common marjoram, Origanum majorana L. 743.
mallikā Arabian jasmin, Jasminum sambac (L.) Ait. 390.
mātulaṅga citron tree, Citrus medica L. 397, 732 = *jambīra.*
mālatī jasmine, Jasminum grandiflorum L. 103, 637, 639, 665.
māṣa black gram, Vigna mungo (L.) Hepper 379, 521, 526, 528, 529, 538, 552, 554, 566, 570, 702.
muñja rush or sedge-like grass, Saccharum bengalense Retz. 448.
mudga mung bean, Vigna radiata (L.) Wilczek. 380, 521, 528, 539, 570, 594.
mocā banana, Musa paradisiaca L. 632, 732 = *kadalī, rambhā.*
yava barley, Hordeum vulgare L. emend. Bourden 367, 577.
yavara millet, Sorghum bicolor (L.) Mönch. 381, 540, 576, 577.
rathadruma Dalbergia ougeinensis Roxb. 749.
rambhā banana, Musa paradisiaca L. 632, 732 = *kadalī, mocā.*
rasāla mango, Mangifera indica L. 660 = *rasālaka* 102, 395, 653, 659 = *āmra.*
rāśijaṭikā Problematic. According to the copyist of the ms. it is identical with Tamil *kottavarai,* closter bean, Cyamopsis psoraliodes (Vacek), however, this identification based on the interpretation of *jaṭikā* is unsound. 384, 604.
lavaṅga the clove tree, Syzygium aromaticum (L.) Merr. et L. M. Perry 671.
laśuna garlic, Allium sativum L. 784.
*likuca*lakoocha, Artocarpus lakoocha Roxb. 392, 632, 651, 660, 745,
vaṁśa bamboo, Bambusa arundinacea (Retz.) Willd. 397, 748.
vaṭa banyan, Indian fig tree, Ficus bengalensis L. 91, 95, 395, 743.
vanavallikā a wild variety of *vallikā.* According to the copyist of the ms. it is identical with Tamil *kāttuppākal* prickly carolah (Vacek). 384, 604.
varṇaja sandal, Santalum album L. 747 = *candana*
vallī creeper. 620, 636, 645, 666, 732 = *vallikā* 384, 604, 626, 627, 671.
vāṭapota the tree Butea frondosa Koen ex Roxb. 98.
vārtāka egg-plant, Solanum melongena L. 384, 604, 620, 621, 626, 639, 641, 645, 651, 657, 666.
vibhītaka belleric myrobalan, Terminalia bellirica (Gaertn.) Roxb. 392, 744.
veṇuka reed 92.
vetasa reed, Calamus rotang L. 91.
vrīhi rice. Oryza sativa L. 363, 365, 366, 367, 368, 425, 497, 560 = *kalama, dhānya, śambaka, śāli.*

śama Prosopis spicigera L. 743.
śami Prosopis cineraria Druce. 391.
śambaka rice, Oryza sativa = *kalama, dhānya, vrīhi, śāli* 363, 364, 458.
śākuṭa Streblus asper Lour. 387, 605, 628, 646.
śāli rice, Oryza sativa, 358, 362, 424, 451, 452, 454, 602. = *kalama, dhānya, vrīhi, śambaka.*
śigru drumstick tree, Moringa oleifera Lam. 391, 743, 785.
śiriṣa lebbeck tree, kokko, Albizia lebbeck (L.) Benth. 393, 747.
śuṇṭhī ginger, Zingiber officinale Roscoe. 387, 606 = *ārdra, ārdraka.*
śyāma lit. 'dark-coloured'. It is the name of 40 different plants (Meulenbeld 1974, 605 ff. and Das 1988, 240) 381.
Śrīparṇa Premna latifolia Roxb. 393 = *śrīparṇaka* 743.
ṣavaka Problematic. The copyist of the ms. identifies it with Tamil *veṇṭai* ladyfinger, Hibiscus esculentus L. 385, 604, 620, 645, 657.
Ṣaṣṭika a kind of rice which ripes in sixty days, Oryza sativa L. 358.
saptaparṇa dita bark tree, Alstonia scholaris (L.) R. Br. 394.
sarjaka myrobalan, Terminalia tomentosa Bedd. = Wight. Et Arn. 250, 396, 742.
sarṣapa mustard, Brassica juncea (L.) Czern. 382, 541, 570, 702.
sāra sal, Shorea robusta Gaertn. F. 730.
sinduvāra East Indian walnut, Vitex trifolia L. 393, 747.
sūraṇa elephant-foot, yam (Randhawa 1980, 501) Amorphophallus campanulatus (Roxb.) Bl. Ex Decne. 387, 605, 628, 636, 646 = *kanda.*
haridrā turmeric, Curcuma longa L. 387, 606, 629, 646.
haimavatī yellow myrobalan, Terminalia chebula Retz. 744.

Bibliography and Abbreviations

AAI Agriculture in ancient India. Ed. D. Raghavan. New Delhi 1964.

Adiceam 1966 E. Adiceam: La géographie de l'irrigation dans le Tamilnad. Paris 1966.

AgP Agnipurāṇa. Ed. by A. B. Upadhyaya. Varanasi 1966.

Agrawala 1974 see BKS

AirGr II 2 A. Debrunner: Altindische Grammatik. II 2. Göttingen 1954.

Ak. Amarakośa of Amarasiṁha. Ed. by N. G. Sardesai and D. G. Padhye. Poona 1969²

AOH Acta Orientalia Academiae Scientiarum Hungaricae.

Ap. Bhuvanadeva: Aparājitapr̥cchā. Ed. P. A. Mankad. Baroda 1950.

ĀpDhS Āpastambadharmasūtra. See Dharmasūtras.

Apte 1957 V.S. Apte's The practical Sanskrit-English dictionary. Poona 1957.

Basham 1981 A. L. Basham: The wonder that was India. Calcutta-Allahabad-Bombay-Delhi 1981. Rupa paperback.

BauDhS Baudhāyanadharmasūtra. See Dharmasūtras.

Beskrovniy 1972 Hindi-Russkiy slovar'. Ed. V.M. Beskrovniy. Moskva 1972.

BhāP Bhāgavatapurāṇa. With Sanskrit text and English translation by C. L. Goswami and M. A. Śāstrī . I –II. Gorakhpur 1971.

Bhargava 1973 Bhargava's Standard illustrated dictionary of the Hindi language. Compiled and edited by R. C. Pathak. Varanasi 1973.

Bhattacharji 1978 S. Bhattacharji: The Indian Theogony. A Comparative study of Indian mythology from the Vedas to the Purāṇas. Calcutta 1978. Corrected Indian edition.

BHS D F. Edgerton: Buddhist hybrid Sanskrit grammar and dictionary. II. Dictionary. First Indian edition. Delhi-Patna-Varanasi 1970.

BhāVyā Bhāṣāvyākhyāna, a commentary in Malayalam on KA. I quote it on the ground of Kangle's *apparatus criticus* to KA.

Bhok. Bhojanakutūhala of Raghunātha. Part I. Published by Śūranād Kunjan Pillai. Trivandrum 1956.

BKS Budhasvāmin: Br̥hatkathāślokasaṁgraha. Ed. P. K. Agrawala. Varanasi 1974.

Brown 1980 C.P. Brown: Dictionary Telugu-English. New Delhi 1980².

BPs Br̥hatparāśarasmr̥ti. In: Dharmaśāstrasaṁgraha II. Ed. by Jībananda Vidyāsāgara. Calcutta 1876.

Bs. Varāhamihira: Br̥hatsaṁhitā with Bhaṭṭotpala's commentary. Ed. A. Tripathi. Varanasi 1968.

BSOAS Bulletin of the School of Oriental and African Studies.

Burrow 1937 T. Burrow: The language of the Kharoṣṭhi documents from Chinese Turkestan. Cambridge 1937.

Caland 1929 Vaikhānasasmārtasūtram. Translated by W. Caland. Calcutta 1929.

CCDPL A comprehensive and critical dictionary of the Prakrit languages. Ed. A. M. Ghatage. Poona 1993–

Chattopadhyaya 1968 D. Chattopadhyaya: Lokāyata. A study in ancient Indian materialism. New Delhi 1968.

Chaudhuri 1980 N. C. Chaudhuri: Hinduism. A religion to live. Oxford – New York-Toronto-Melbourne 1980.

Choudhary 1971 A. K. Choudhary: Early medieval village in North-Eastern India. A.D. 600 – 1200. Calcutta 1971.

Colas Personal communication by Gerard Colas (Paris)

Crooke 1926 W. Crooke: Religion & Folklore of Northern India. Prepared for the press by R. E. Enthoven. Oxford 1926.

Das 2001 R. P. Das: Bemerkungen anlässlich einer neuen Arbeit zum *Pākadarpaṇa*. In: TSAM 6 (2001), 120–135.

DKC The Daśakumāracarita of Daṇḍin. Ed. and transl. by M. R. Kale. Delhi-Varanasi-Patna 1966[4].

Datta 1989 Swati Datta (née Sen Gupta): Migrant Brāhmaṇas in Northern India. Delhi-Varanasi-Patna etc. 1989.

Dātāra Kauṭalīyamarthaśastram of Ācārya Viṣṇugupta with four commentaries. Ed. V. Ś. Dātāra. Varanasi 1991.

Dharmasūtras The law codes of Āpastamba, Gautama, Baudhāyana and Vasiṣṭha (annotated text and translation) by P. Olivelle. Delhi 2000.

EDSHP An encyclopaedic dictionary of Sanskrit on historical principles. Ed. A. M. Ghatage and others Poona 1976–

EWA M. Mayrhofer: Kurzgefasstes etymologisches Wörterbuch des Altindoarischen. Heidelberg 1986–2001.

GDhS Gautamadharmasūtra. See Dharmasūtras.

Glasenapp 1922 H. v. Glasenapp: Der Hinduismus. Religion und Gesellschaft im heutigen Indien. München 1922.

Gode 1943 P.K. Gode: Some notes on the history of Indian dietetics with special reference to the history of *jalebī*. In: NIA 6: 8–9. November-December 1943, 169–181.

Gode 1969 P. K. Gode: The use of *kāca* or glass in Indian pharmacy. In: Studies in Indian cultural history III. Poona 1969, 89–101.

Gonda 1970 J. Gonda: Viṣṇuism and Śaivism. A comparison. London 1970. (Jordan lectures 1969)

Gonda 1977 J. Gonda: Medieval religious literature in Sanskrit. Wiesbaden 1977. (A History of Indian Literature II. 1.)

Gonda 1980 J. Gonda: Vedic ritual. The non-solemn-rites. Leiden – Köln 1980.

Gonda 1993 J. Gonda: Aspects of early Viṣṇuism. New Delhi 1993. (First edition: 1954)

Gopal 1962 L. Gopal: The Śukranīti-a nineteenth century text. In: BSOAS 25: 1–3 (1962), 524–556.

Goudriaan Personal communication by Teun Goudraan (Utrecht)

Goudriaan 1965 Kāśyapajñānakāṇḍa. Kāśyapa's book of wisdom. A ritual handbook of the Vaikhānasas. Translated and annotated by T. Goudriaan. The Hague 1965.

Grierson 1985 G.A. Grierson: Bihar peasant life. Reprint edition. New Delhi 1985.

GRK Gr̥hastha-Ratnākara by Caṇḍeśvara Ṭhakkura. Rd. By Mahāmahopādhyāya Kamalakr̥ṣṇa Smr̥titīrtha. Calcutta 1928.

IIJ Indo-Iranian Journal

Jm. Jayamaṅgalā of Yaśodhara. See Śrīvātsyāyanamaharṣipraṇītaṁ kāmasūtram yaśodharaviracitayā jayamaṅgalākhyavyākhyayā sahitam. Kalyāṇa-Baṁbaī. Saṁvat 1991, Śake 1856. [1934–5].

JESHO Journal of Economic and Social History of Orient

KA The Kauṭilīya Arthaśāstra. Part I. A critical edition with a glossary by R. P. Kangle. Bombay 1969.

KN Kāmandakīyanītisāra. Ed. by J. P. Miśra. Bambaī. Saṁvat 2009, śake 1874 [1952].

Kane 1968 P.V. Kane: History of Dharmaśāstra. I./I.. Revised and enlarged. Poona 1968.

Kane 1974 P.V. Kane: History of Dharmaśāstra. II./II. Revised and enlarged. Poona 1974.

Kangle 1972 The Kauṭilīya Arthaśāstra. Part II. An English translation with critical and explanatory notes. Bombay 1972².

KāS Vr̥ddhajīvaka: Kāśyapasaṁhitā. Ed. V.J.T.Ācārya and S. Śarmā. [Muṁbaī] 1938.

KāKS Kāśyapīyakr̥ṣisūkti

Kosambi 1975 D.D. Kosambi: An introduction to the study of Indian history. Bombay 1975².

Kölver 1988 B. Kölver: On the origins of the jajmānī system. In: JESHO 31:1 (1988), 265–285.

Kr̥G Kr̥ṣigītā. in: Pāṭṭukal II. Ed. R. Pillai. Trivandrum 1968, 109–142.

Kr̥K Kr̥ṣikoś I-II. Saṁpādak Viśvanāth Prasād. Paṭnā 1959–1966.

Kr̥P Kr̥siparāśara. Ed. G. Majumdar and S. C. Banerji. Calcutta 1960.

KŚ Kāśyapaśilpa. Ed. by V. G. Apte. Poona 1926.

KSS Somadeva: Kathāsaritsāgara. Ed. J. Śāstri. Delhi-Patna-Varanasi 1970.

Ks. Kumārasambhavam or The birth of war-god. Text with English Translation by H.H. Wilson. Varanasi 1966.

Km. Kuttani-Matam or Shambhali-Matam by Damodara Gupta. Ed. by T. N. Tripathi. Varanasi 1991².

MahāBhā Mahābhārata. Ed. V. S. Sukthankar and others. Poona 1927–1966.

MaS Manu's code of law. A critical edition and translation of the Mānava – Dharmaśāstra by P. Olivelle. Oxford 2005.

Mās. Mānasāra. Ed. by P. K. Āchārya. London 1933.

Mav. Mahāvyutpatti. Ed. R. Sakaki. Reprint edition. Kyoto 1962, 1965.

Mahias 1988 M.C-Mahias: Milk and its transmutations in Indian society. In: Food and Foodways. 2 (1988), 265–288.

Majumdar 1971 R. C. Majumdar: Medicine. in: A concise history of science in India. Ed. D.M. Bose-S.N. Sen and B.V. Subbarayappa. New Delhi 1971, 213–268.

Meulenbeld Personal communication by Prof. G. J. Meulenbeld.

Meulenbeld 1984 G. J. Meulenbeld: „The Surveying of Sanskrit medical literature". In: Rijksuniversiteit te Groningen, Publikaties van het Instituut voor indische telen en culturen 4, 1984, 31–114.

Meyer 1926 Das altindische Buch vom Welt- und Staatsleben. Das Arthaṣāstra des Kauṭilya. Aus dem Sanskrit übersetzt und mit Einleitung und Anmerkungen versehen von Johann Jakob Meyer. Leipzig 1926.

Meyer 1937 J. J. Meyer: Trilogie der altindischer Mächte und Feste der Vegetation. Zürich-Leipzig 1937.

MW M. Monier-Williams: A Sanskrit-English dictionary. Oxford 1960.

Mylius 1974 K. Mylius: Das altindische Ernteopfer, der *āgrayaṇa*. In: Asien in Vergangenheit und Gegenwart. Berlin 1974, 433 – 442. (Beiträge der Asienwissenschaftler der DDR zum XXIX. Internationalen Orientalistenkongress 1973 in Paris)

Myrdal-Runebjer 1988 E. Myrdal-Runebjer: Report spring 1988. Development of agricultural and live-stock tending implements and techniques in Tamil Nadu, after the introduction of iron to c. A.D. 1800: a choise of perspective. Typewriting, 40 p.

NāS Nāradasmṛti. Ed. R. Lariviere. Philadelphia 1989.
NCC New Catalogus Catalogorum IV. Ed. V. Ragvhavan. Madras 1968.
NIA New Indian Antiquary.
NīP Nīlamatapurāṇa. Ed. Ved Kumari. Srinagar 1973.
Nyammasch 1984 M. Nyammasch: Untersuchung zur Genesis des Feudalismus in Indien. Berlin 1984.
Olivelle 2002 P. Olivelle: Food and Thought. Dietary rules and social organization in ancient India. Amsterdam 2002. (Gonda lectures)
Olivelle 2004 see MaS
Pandey 1969 R. Pandey:Hindu Saṁskāras. Delhi 1969.
P Pāṇini: Aṣṭādhyāyi. Hrsg, O. Böhtlingk. Leipzig 1887.
PaP Padmapurāṇa. Ed. by Viśvanātha Nārāyaṇa Maṇḍalīka. Bombay 1893–1895.
PāGS Pāraskaragṛhyasūtra. M. G. Bākre. Bombay 1917.
Pd. Pākadarpaṇam by Mahārāja Nala. Ed. by Sri Vamacharana Bhattacharya. Benares 1915.
PaŚR Dhanañjayabhaṭṭa: Paryāyaśabdaratna. Ed. E.D. Kulkarni and M. C. Dixit. Poona 1971.
Payne 1933 E. A. Payne: The Śāktas: an introductory study. Calcutta 1933.
Pratipadapañcikā. See Dātāra.
PTSD T.W. Rhys Davids- W. Stede: The Pali Text Society's Pali-English dictionary. London 1966.
Rv. Raghuvaṁśa of Kālidāsa. With the commentary of Mallinātha and Hindi translation by Haragovinda Miśra. Vārāṇasī saṁvat 2034 [1977]
Rt. Kalhaṇa: Rājataraṅgiṇī. Ed. M.A.Stein. Bombay 1892.
Rām. Vālmiki: Rāmāyaṇa. Ed. P. R. Śāstri. Kāśi 1959.
Randhawa 1980 M. S. Randhawa: A history of agriculture in India. I. Beginning to 12th century. New Delhi 1980.
Randhawa 1982 M. S. Randhawa: A history of agriculture in India. II. Eight to eighteenth century. New Delhi 1982.
Rao 1985 T.A.G. Rao: Elements of Hindu iconography. Delhi-Varanasi-Patna-Madras 1985^2.
Rau 1983 W. Rau: Die Brennlinse im alten Indien. Wiesbaden 1983.
RV Rig Veda. A metrically restored text. Ed. by B. A. van Nooten and G. B. Holland. Cambridge Massachusetts 1964.
RVD W.D. Whitney: The roots, verb-forms, and primary derivatives of the Sanskrit language. Reprint edition. Delhi 1979.
Sangwan 1990 S. Sangwan: European impressions on science and technology in India. In: History of Science and Technology in India. Ed. by G. Kuppuram K. Kumudamani Delhi 1990, V, 65–101.
SE Kashyapiyakrishisukti (A treatise on agriculture by Kashyapa) Translated by S.M. Ayachit. Commentaries by Nalini Sadhale and Y.L. Nene. Secunderabad 2002. (Agri-History Bulletin 4.)
Sen 1978 C. Sen: A dictionary of the Vedic rituals. Delhi 1978.
Sen Gupta 1956 S. Sen Gupta: Food prohibition in smṛti texts. In: Journal of the Asiatic Society Letters XXII/ 2. 1956, 169–209.
Sharma 2003 A. Sharma: Hinduism and its sense of history. Oxford 2003.
Sharma Personal communication by R. S. Sharma (Delhi)

Sharma 1980 R. S. Sharma: Śūdras in ancient India.Delhi-Varanasi-Patna 1980[2].

Singh 1968 R.C.P. Singh: Kingship in Northern India (Cir. 600A.D.-1200 A.D.). Delhi-Varanasi-Patna 1968.

ŚRK Śivatattvaratnākara of Basavarāja. I – II. Mysore 1964 – 1969.

Śrīmūla Śrīmūla of Mahāmahopādhyāya T. Gaṇapati Śāstrī on KA. See Dātāra.

Stein 1994 B.Stein: Peasant state and society in medieval South India. Oxford India paperback. Delhi-Oxford-New York 1994.

Stevenson 1971 S. Stevenson: The rites of the twice – born. New Delhi 1971[2].

Stutley and Stutley 1977 Margaret and James Stuttley: A dictionary of Hinduism. Its mythology, folklore and development 1500 B.C. – A.D. 1500. London 1977.

Thapar 2002 R. Thapar: Early India. From the origins to AD 1300. London 2002.

TSAM Traditional South Asian Medicine

Unni 2006 The Arthaśāstra of Kauṭalya. English translation by N. P. Unni. Delhi 2006.

Uv. G.P. Majumdar: Upavana-vinoda (A Sanskrit treatise on arbori-horticulture). Calcutta 1935.

Vacek Personal communication by Prof. Jaroslav Vacek (Prague)

VaiSmS Vaikhānasasmārtasūtra. Ed. W. Caland. Calcutta 1927.

VāsDhS Vāsiṣṭhadharmasūtra see Dharmasūtras.

Vś Vairāgyaśataka. See Love lyrics by Amaru & Bhartri-hari. Translated by G. Bailey. New York 2005. (Clay Sanskrit Library)

ViS Viṣṇusmṛti. Ed. by J. Jolly. Varanasi 1962.

Vv. Viśvavallabha. See Vishvavallabha (Dear to the world: the science of plant life). Transl. Nalini Sadhale. Commentaries by Nalini Sadhale and Y.L. Nene. Secunderabad 2004. (Agri-History Bulletin 5.)

Vā Sūrapāla: Vṛkṣāyurveda. Hrsg. R. P. Das. Stuttgart 1988.

Watters 1973 On Yuan Chwang's Travels in India. Translated by T. Watters. New Delhi 1973[2].

Whitehead 1921 H. Whitehead: The village gods of Sout India. London – New York – Toronto etc. 1921.

Winternitz 1967 M. Winternitz: History of Indian literature. II 2 (Scientific literature). Delhi-Varanasi-Patna 1967.

Wojtilla 1985 Gy. Wojtilla: Skt. kheṭa-. In: IIJ 28: 3 (1985), 200.

Wojtilla 1995 Gy. Wojtilla: Some remarks on the Kāśyapīyakṛṣisūkti. in: International conference on Sanskrit and related studies. September 23–26, 1993. (Proceedings). Ed. C. Galewicz. Cracow 1995, 269–274.

Wojtilla 2003 Gy. Wojtilla: Etymological miscellany. In: Sarūpa-Saurabham. Tributes to Indology. Prof. Lakshman Sarup Centenary Volume. Ed. Ashvini Agrawal. New Delhi 2003, 41–45.

Wojtilla 2004 Gy. Wojtilla: Notes on pākaśāstra, sūdaśāstra and sūpaśāstra. In: Du corps humain, au carrefour de plusieurs savoirs en Inde. Mélanges offerts à Arion Roçu par ses collègues et ses amis à l'occasion de son 80e anniversaire. Ed. E. Ciurtin. Bucarest-Paris 2004, 337–346. (Studia Asiatica IV-V.)

Wojtilla 2005a Gy. Wojtilla: The Sītādhyakṣaprakaraṇa of the Arthaśāstra. In: Indische Kultur im Kontext. Rituale, Texte und Ideen aus Indien und der Welt. Festschrift für Klaus Mylius. Hrsg. von L. Göhler. Wiesbaden 2005, 413–425.

Wojtilla 2005b Gy. Wojtilla: A preliminary report on the *Kṛṣisamayaniṇaya.* In: Encyclopedia of Indian Wisdom. The Felicitation Volume of Prof. Satya Vrat Shastri. Delhi 2005, 375–380.

Wojtilla 2005c Gy. Wojtilla: A forgotten Sanskrit text on post-Arthaśāstra economic policy. In: AA 45 (2005), 343–351.

Wojtilla 2006 Gy.Wojtilla: History of Kṛṣiśāstra. Wiesbaden 2006. (Beiträge zur Kenntnis südasiatischer Sprachen und Literaturen 14)

YāS Yājñavalkyasmṛti with the commentary Mitākṣarā of Vijñāneśvara. Ed by Narayan Ram Acharya. Bombay 1949[5].

Yt. Yaśastilakacampū of Somadeva Sūri. Ed. by Śivadatta, Kāśināth Pāṇḍuraṅg Parab and Vāsudeva Laxmaṇ Śāstrī Paṇaśikar. Bombay 1903–1916.